BASIC Geography

Gill Rollins

Richard Kemp   Editorial Adviser

MACMILLAN

Acknowledgements

The author and publishers wish to thank the following who have kindly given permission for the use of copyright material:

Bishops Bowl Lakes
Cunard
Daily Mail
Department of Transport (Schematic Motorway Map)
Financial Times
George Philip & Son Limited
The Independent
Intercity (portion of map)
Leeds University
London Country North West Green Line
The Open University
Whipsnade Park Zoo

The author and publishers wish to acknowledge, with thanks, the following photographic sources:

Aerofilms
Science Photo Library

The publishers have made every effort to trace the copyright holders, but if they have inadvertently overlooked any, they will be pleased to make the necessary arrangements at the first opportunity.

First published 1988
Reprinted 1989, 1990

Published by
MACMILLAN EDUCATION LTD
Houndmills, Basingstoke, Hampshire RG21 2XS
and London
Companies and representatives
throughout the world

Designed, typeset and illustrated by
Gecko Limited, Bicester, Oxon

Produced by AMR for
Macmillan Education Ltd

Printed in Hong Kong

British Library Cataloguing in Publication Data
Rollins, Gill
Basic Geography Gill Rollins.—(Basic tests).
1. Geography—For schools
I. Title II. Series
910
ISBN 0–333–45137–6

Contents

Introduction

This book forms part of a Macmillan series designed for use by a wide range of students in school and further education, to provide practice in the development of core skills and knowledge in a variety of essential areas.

In particular, the series aims to ensure competence in core skills so that potential employees – school and college leavers – actually master the basic skills required by employers. Many of the skills included in the series are also needed in everyday life.

The approach taken is to present information in clear and carefully controlled steps and to provide numerous straightforward questions and tasks designed to test skills and explore the information presented.

This book is targeted at the A.E.B. Basic Geography Test and can also be used to help students prepare for the new Basic Test (Special) in Geography for Tourism and Leisure.

In clear, simple terms it helps students to acquire a working knowledge and understanding of United Kingdom, European and world locations: several organisations have commented recently that their young employees do not have this basic geographical knowledge.

It also encourages the development of practical mapwork skills that can be useful to students in everyday life, such as when finding their way in cities or planning journeys of all kinds.

Each part of the first three sections (on Mapwork Skills, UK Locations and World Locations) is designed as a self-contained double page unit, with information on one side and exercises (based largely on real life situations) on the other.

The final section, on solving problems in geography, gives pupils the opportunity to apply their skills and knowledge in a series of assignments set in the UK and other parts of the world.

Mapwork Skills

Map symbols and map keys

Before people knew how to draw maps as we know them today, they would draw picture maps to describe places.

This old picture map tells us that there are hills, a town, some woods and some lines that might be rivers or roads.

Old map

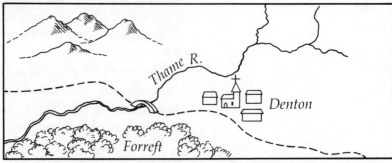

Today, we would draw a map using a *scale* (which you will learn about on page 12) and *symbols*. These tell us *exactly* what is shown on the map. To understand the symbol, we need a *key*, which would look like this:

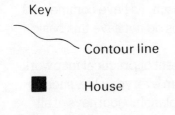

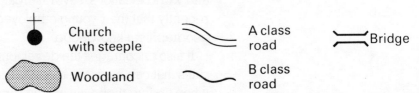

Using the key, we can now understand much more about the area shown on the old map above.

Modern Map

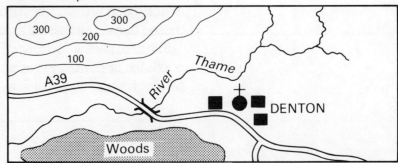

How much more does this map tell us about the hills, for instance how high they are? How many houses are there in the village? What sort of church is shown? What types of road are there in this area?

You will be looking at many different kinds of map in this book. They will all use symbols and they will all have keys.

Here are some examples of symbols that you may know already:

How many more symbols can you draw that are in everyday use?

Tasks

1 Look at this map of Whipsnade Zoo, and use the key to answer the questions.

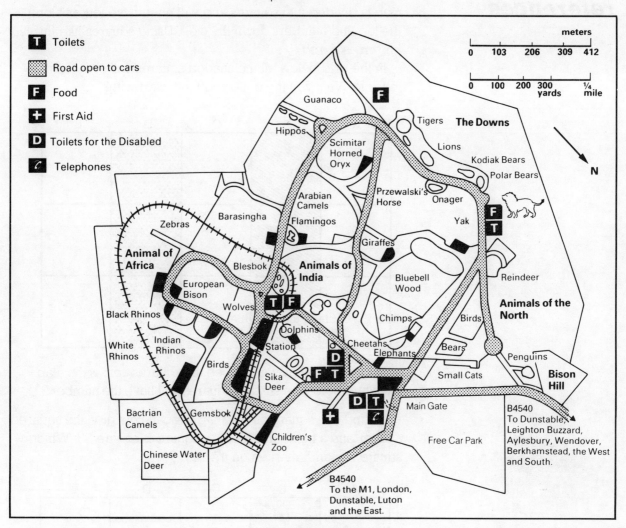

Key:
- **T** Toilets
- Road open to cars
- **F** Food
- **+** First Aid
- **D** Toilets for the Disabled
- **☎** Telephones

a If you came by car from London, which road would bring you to the Main Gate?

b Could you drive your car to see the Chinese Water Deer?

c If someone needed to make an urgent phone call, how would you describe to them where the telephones are?

d How many toilets are there for the disabled?

e Which would be the best way to see the European Bison? By train or by road?

f How many places provide food at Whipsnade?

g Which small animals are near the Main Gate?

h Find the Children's Zoo and Bison Hill. Which is nearest to the First Aid Post?

2 Look at an Ordnance Survey 1:50 000 map (use one of your own area if you can). Use the Key to find out what these symbols show:

3 Find a local map of your area, such as a town centre plan. Draw some of the symbols used (you will need to use the Key). Now work with a friend, and test each other on the symbols shown on the map.

Grid and square references

When we look at a map we need a way to find (or to tell other people how to find) places quickly. To help us to do this, the map is divided into squares and will often have letters along the top and numbers down the side. Using squares like this is known as a *grid*.

In the grid below all the shaded squares will start with the letter B. What will the dotted squares start with?

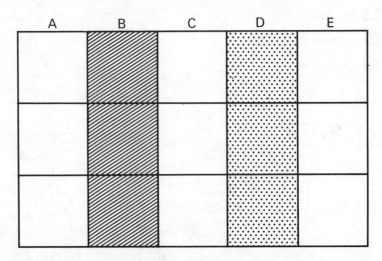

In the grid below you can see the squares *across* the map have numbers. The shaded squares will have the number 2.

To find one square on the map we can now give the square a letter *and* a number. The dark square is square B3. Which square has an X marked on it?

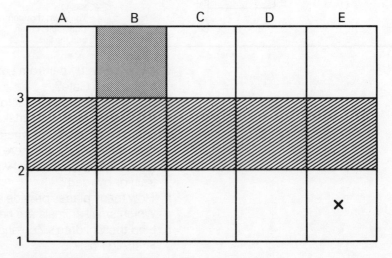

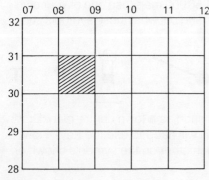

Number grids

On some grids, numbers only are used. Follow the same rule: Give the numbers along the *top* first and you will be able to find the right square. The shaded square is 0830. What number is the square in the top right hand corner?

You will notice two things about these numbers:
1 There are *no gaps* between them.
2 There are always *four* numbers. Single numbers, like 8, will have an 0 in front, to become 08.

Tasks

1 Look at the map of Lagdon Peak and answer these questions:
a Steve, walking in the hills, falls and breaks his leg at point X on the map. His friend Linda has to find the nearest telephone, to dial for an ambulance. Which square is the telephone in?
b How does Linda describe where Steve can be found? Which square is he in? What landmark on the map is he near?
c Which square is the hospital in?
d What road would the ambulance use to get to the accident?
e Would the ambulance be able to drive right up to Steve?
f If Steve is trapped under some rocks, what building in square E8 will Linda need to phone as well?

Lagdon Peak

Key

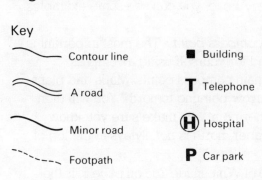

Contour line —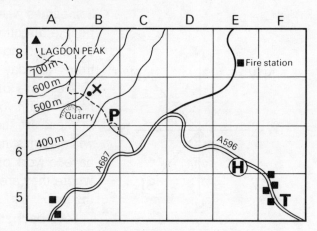
A road
Minor road
Footpath

■ Building
T Telephone
H Hospital
P Car park

2 Now study carefully this map of a part of London.

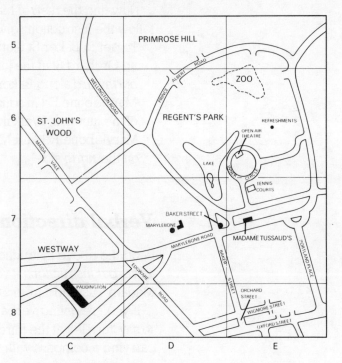

a What two stations are in square D7?
b Which square is Madame Tussaud's in? (a tourist attraction).
c Find the road called the Inner Circle in square E6. What sort of theatre is on this road?
d What sport could you play in square E7?
e In which square in the park could you buy some refreshments?

3 Using the map of your local area find your own home, and give its square reference. What other useful places can you find by using square references?

Giving directions

Compass directions

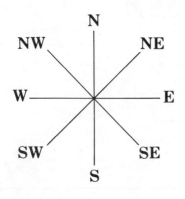

Sometimes when we want to find places on maps, or to give people directions for journeys, we need to know where north, south, east and west are. Above, you can see some examples from everyday life.

These are known as *compass points*. The most important is north, which you can find for yourself using a compass.

These are the eight main compass points. Maps and plans will often just have an arrow pointing to north. You will then have to work out the other points, so make sure you know where they are. Remember, north is *not* always at the top of the map.

Look back at the map of Whipsnade Zoo on page 7. Is the Car Park on the *north* or the *south* of the map?

Now find the map of Marylebone on page 9. See if you can follow the instructions given below.

To get to Baker St Station from Oxford St, turn *north* up Orchard St at the junction in square E8. Then continue *northward* along Baker St until you reach the junction with Marylebone Rd in square D7. The station is on the *east* side of the junction. Walk in an *easterly* direction along Marylebone Rd and Madam Tussaud's is a little way along on the *north* side.

Verbal directions

Read the above instructions. You have given what are known as *verbal instructions* – you have *told* someone how to get to Madam Tussaud's from Oxford Street, via Baker Street Station.

If you are giving verbal directions without a map, it is easier to say 'turn left at the third crossroads'. Not many people will be carrying a compass with them!

Tasks

1 Think of a destination (the place at the end of a journey) that can be reached easily from your school or college. Tell a friend how to get there on foot, speaking slowly and carefully. See if your friend can guess the destination. Try tape recording your directions, and see if *you* can understand them!

2 Now look at the map on page 11, which shows the centre of Leeds. You can do this work on your own or in a group.

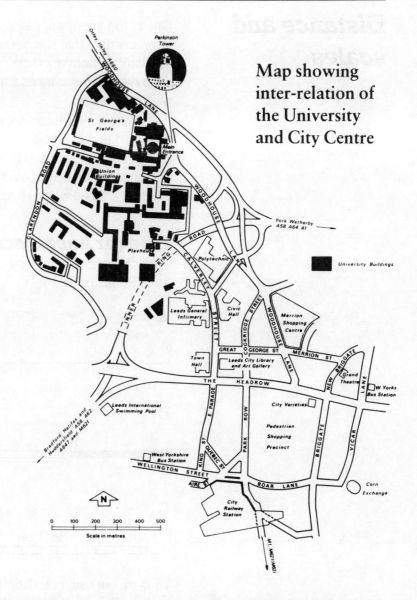

Map showing inter-relation of the University and City Centre

a If you are standing in Calverley Street outside the Civic Hall and facing *north*, is the Leeds General Infirmary on your left or your right?

b Is the City Railway Station *south-west* or *south-east* of the Pedestrian Shopping Precinct?

c What is the destination of this journey?
Start at the City Railway Station, and walk *north* up Park Row until you reach the junction with the Headrow. Turn *east* and then take the second turning to the *north*. Your destination is a little way along on the *east* side of New Briggate.

d Now rewrite these instructions, using *left* and *right* instead of compass directions.

e Work with a friend, each looking at the map. Describe these journeys (planning them first and without revealing the destination) and see if your friend can find the correct destination.
– From the West Yorks Bus Station to the Leeds International Swimming Pool.
– From the Parkinson Tower to the Town Hall.

3 Look back to the map on page 9. How would Linda tell the ambulance driver to find Steve? What directions would she give if she phones the fire station?

Distance and scales

One of the important pieces of information you can find out from any map that has been accurately drawn is the *distance* one thing is from another.

To measure distances you need to use the *scale* on that particular map.

Just like a model car is a scaled down version of the real car, so a map is a scaled down version of what is actually found on the ground. Using the map and its scale, you can work out what the actual distances would be on the ground shown by the map. There are three ways a map's scale can be shown.

Line (or linear) scales

This is the most usual type of scale, and the sort you will be using in this book.

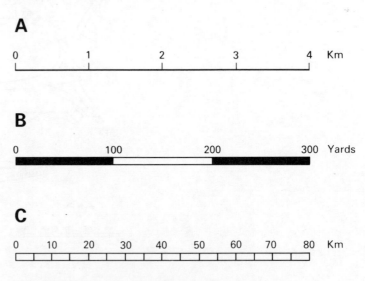

To work out distances using a line scale:
1 Use a ruler to measure the distance on the map.
2 Place your ruler against the line scale. Read off the actual distance it would represent on the ground.

Written scales

Line scale A (above) tells us that *2 centimetres* on the map represents *1 kilometre* on the actual ground.

Line scale B tells us that *1 inch* on the map represents *100 yards* on the actual ground.

Fraction scales

Some maps and plans (OS maps for example) also write the scale as a fraction.

The fraction scale for line scale A would be 1:50000. This means that *1 centimetre* on the map represents *50000 centimetres* on the actual ground.

Fraction scales will not be used in this book, so this is just to explain what they are if you come across them.

Tasks

1 What would be the written scale for *line scale C* shown on the opposite page?

2 Draw your own line scale for these written scales:
   a 1 inch represents 1 mile.
   b 1 centimetre represents 10 kilometres.
   c ½ inch represents 10 miles.
   d 2 centimetres represents 10 metres.

3 Using *line scale C* again, measure the distances between all these places, and fill in your results in your copy of the table.

● NORTHVILLE

EASTHAM ●

● WESTBURY

● SOUTHTON

NORTHVILLE – EASTHAM	km
NORTHVILLE – WESTBURY	km
NORTHVILLE – SOUTHTON	km
EASTHAM – WESTBURY	km
EASTHAM – SOUTHTON	km
WESTBURY – SOUTHTON	km

4 The map below shows the route for a Charity Fun Run, which you and friends are going to take part in. Using the map work out:
   a How far you have to go before the *first* Drinks Stop.
   b How far it is between the *first* and *second* Drinks Stops.
   c The total distance of the Fun Run course.
   d What would you be passing after running 5 km and 12½ km?
   e In what compass direction would you be running on each leg of the run?

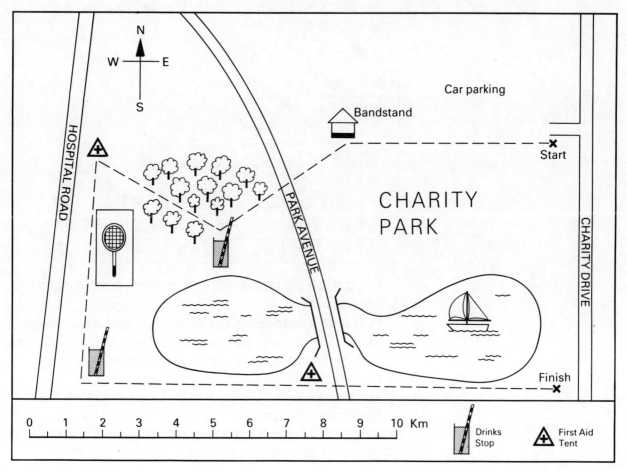

Time-distance and distance charts

Time-distance

When you ask friends how far they are from school, they might well say 'twenty minutes by bus'. They would be very unlikely to give you an exact answer in miles or kilometres.

This is because the *time* a journey takes is often more important to people than the distance travelled.

Below are two examples of 'time-distance' as this way of describing journeys is known. Try and think of your own.

'FLY THE UNDERGROUND – HEATHROW IN 40 MINUTES'
'NEW FLATS FOR SALE – 5 MINUTES FROM SHOPS, 10 MINUTES FROM STATION'

Why time is important

Some people (such as ambulance drivers carrying urgent blood supplies) need to make journeys very quickly. Can you think of some journeys where saving time is important?

Planning journeys

When you plan a journey (either for yourself or for other people), you will need to know the actual distance. Then you can work out how much time the journey will take.

Distance charts

Distance charts can help you to do this. They are found in atlases, guidebooks and motoring handbooks. (See if you can find one in school or at home).

This distance chart helps you to find the distance between cities in England and Wales:-

B'ham	Bristol	Edinburgh	Glasgow	Leeds	Liverpool	Manchester	Newcastle	Sheffield
85								
286	365							
287	365	44						
109	194	191	210					
90	160	210	212	33				
79	159	210	211	40	35			
200	284	106	143	91	153	128		
77	163	223	240	33	73	38	123	

You can use the chart to find the distance from Birmingham to Leeds like this:

1 Find Birmingham on the chart. The numbers on the column (going down the page) are the distances from Birmingham to the other towns. Run your finger down the column.
2 Now find Leeds on the chart.
3 Run your finger across the numbers until you reach the Birmingham column.

You have now found the distance you want: the answer is 109 miles. Practise using this chart.

Tasks

1 Here are four ways of travelling from London to Paris: the distance in miles for each journey is roughly the same.

Car		Train		Air		Personal helicopter	
Drive London – Dover	2 hr	Victoria – Dover	1½ hr	Taxi London H'row	1 hr	Central London to Central Paris	1 hr
Ferry Dover – Calais	1 hr	Ferry Dover – Calais	1 hr	Heathrow – Le Bourget	½ hr		
Drive Calais – Paris	3 hr	Calais – Gare du Nord	1½ hr	Taxi to Central Paris	1 hr		

  a Which journey takes the longest time? Which is the shortest journey?
  b The helicopter goes more slowly than the passenger jet. Why does the whole journey by taxi and aeroplane take longer?
  c Which of the four journeys would these people be most likely to take?
    i A family setting off for a touring holiday in France.
    ii A businessman making a regular trip to his Head Office in Paris.
    iii A group of young students visiting Paris for the first time.
    iv A pop star rushing to Paris to record a TV programme.

2 You are a sales representative, based in Bristol, and need to plan a journey round England and Wales to sell your company's products. Use the distance chart opposite to complete the route plan below, which your boss must see before you go.

ROUTE PLAN			
Stage of Journey	Starting point	Destination	Distance (miles)
1	Bristol	Birmingham	
2	Birmingham	Manchester	
3	Manchester	Newcastle	
4	Newcastle	Leeds	
5	Leeds	Birmingham	
6	Birmingham	Bristol	

3 When she sees your plan, she reminds you that you must not travel more than 100 miles in one day. Which stages of your journey are too long for one day's travel?

4 When you have finished your journey, you will need to fill in an expenses form, as you will be paid 30 pence for every mile you travel. Fill in this form, to make sure you are paid the correct amount.

Expenses form	
Total distance: miles	
Cost at 30p per mile	£ : p
Signed:	

15

Time-distance and timetables

When you are planning a journey by bus, train or plane, you will have to use a timetable. This will not only tell you how to get to your destination, but also *how long* the journey will take. (Remember what you have learned about time-distance on page 14). You can also work out the details of your journey, such as where to change trains or buses: you must always allow plenty of time for this!

When you arrive at a station or airport, it is important to look at the *arrivals* and *departures* notice boards. They show any delays or cancellations, so that you can change your plans.

How to use a timetable

Some timetables have an index, so that you can look up the places where your journey will begin or end. Sometimes you will need to look through the pages to find your chosen route.

Look carefully at this part of a London Country North West timetable and plan a journey from High Wycombe Bus Station to Amersham Station. It is a Friday morning. You will need to arrivce in time to catch a train at 0830 for London.

Mondays to Fridays

Service:								372 SD	372 SD	NSD											
Marlow Queens Road	....	....	....	....	....	....	....	0747	0749	0749	0813	0833	....	....							
Marlow Bottom Turning	....	....	....	....	....	....	....	0753	0756	0756	0819	0839	....	....							
Cressex Road Marlow Road	....	....	....	....	....	....	....	0758	0802	0802	0824	0844	....	....							
Coronation Road Blenheim Road								↓	↓	↓	↓	↓									
Desborough Avenue Wendover Arms	....	....	....	....	....	**SD**	**NSD**	0801	0805	0805	0827	0847	....								
High Wycombe Bus Station◆	....	....	0654	0732	0757	0802	0809	0813	0813	0835	0852	0905	0935		05	35			1335	1405
Terriers Cross Roads	....	....	0705	0743	0808	0813	0820	0825	0825	0846	....	0916	0946		16	46			1346	1416
Hazlemere Cross Roads	....	0645	0709	0747	0812	0817	0824	0829	0829	0850	....	0920	0950	Then	20	50	U		1350	1420
Holmer Green Bat & Ball	....	0651	0715	0753	0818	0823			0836	0856	....	0926	0956	at	26	56	N		1356	1426
Penn Street	....	0657	0721	0759	0824	0829			0842	0902	....	0932	1002	these	32	02	T		1402	1432
Amersham Bus Garage‡	0613	0630	0647	0705	0731	0809	0834	0839	0837	0842	0852	0912	....	0942	1012	mins	42	12	I	1412	1442
Stanley Hill Brudenell School	↓	↓	↓	↓	↓	0838	↓	0841	0846	↓				↓	↓	past	↓	↓	L	↓	↓
Amersham Station ⊖ ⇌	0617	0634	0651	0710	0736	0814	0844	....	....	....	0917	....	0947	1017	each	47	17		1417	1447
Chesham Broadway ⊖	0626	0643	0700	0721	0747	0825	0855	0855	....	....	0928	....	0958	1028	hour	58	28		1428	1503ε
Chesham Pond Park Estate	0634	0651	0711	0732	0757	0835	0905	0905	....	....	0938	....	1008	1038		08	38		1438	1513

										SD										
Marlow Queens Road	....	....	....				1638	....		1748	1804	....	....							
Marlow Bottom Turning	....	....	..	**SD**		**SD**	1645	....		1755	1810	....	....							
Cressex Road Marlow Road	....	....	1602H		1624H		1651	....		1801	1815	....	....							
Coronation Road Blenheim Road	....						1654	....		1804	1818	....								
Desborough Avenue Wendover Arms	....	**SD**	↓	**NSD**	↓	**NSD**	1659	....		1809	1823	....								
High Wycombe Bus Station◆	1435	1505	1535	1552	1610	1615	1632	1632	1650	1707	1733	1754	1805	1817	1831	1853	1910	1955	2055	2155	2255
Terriers Cross Roads	1446	1516	1546	1604	1621	1626	1643	1643	1701	1719	1744	1805	1816	1829	1842	1904	1919	2004	2104	2204	2304
Hazlemere Cross Roads	1450	1520	1550	1608	1625	1630	1647	1647	1705	1723	1748	1809	1820	1833	1846	1908	1923	2008	2108	2208	2308
Holmer Green Bat & Ball	1456	1526	1556	1615	1631	1636	1653	1653	1711	1730	1754	1815	1826	1840	1852	1913	1928	2013	2113	2213	2313
Penn Street	1502	1532	1602	1621	1637	1642	1659	1659		1736	1800	1821	1832	1846	1858	1919	1934	2019	2119	2219	2319
Amersham Bus Garage‡	1512	1542	1612	1631	1647	1652	1709	1709		1746	1810	1831	1842	1856	1908	1927	1942	2027	2127	2227	2327
Stanley Hill Brudenell School	↓	↓	↓		1651		↓	↓		↓	↓	↓	↓	↓	↓	↓	↓	↓	↓	↓	↓
Amersham Station ⊖ ⇌	1517	1547	1617		1657	1657	1714	1714		1751	1815	1841ε		1901		1931		2031	2131	2231	
Chesham Broadway ⊖	1528	1558	1628		1708	1708	1725	1725		1802	1826	1852		1915ε		1940		2040	2140	2240	
Chesham Pond Park Estate	1538	1608	1638		1718	1718	1735	1735		1813	1836	1902		1924		1948		2048	2148	2248	

Saturdays

High Wycombe Bus Station	....	....	0813	....	0905	0935		05	35			1505	1535	1605	1635	1705	1745	1820	1855	1925	1955
Terriers Cross Roads	....	....	0822	....	0916	0946	Then	16	46			1516	1546	1614	1644	1714	1754	1829	1904	1934	2004
Hazlemere Cross Roads	....	....	0826	....	0920	0950	at	20	50	U		1520	1550	1618	1648	1718	1758	1833	1908	1938	2008
Holmer Green Bat & Ball	....	....	0831	....	0926	0956	these	26	56	N		1526	1556	1623	1653	1723	1803	1838	1913	1943	2013
Penn Street	....	....	0837	....	0932	1002	minutes	32	02	T		1532	1602	1629	1659	1729	1809	1844	1919	1949	2019
Amersham Bus Garage‡	0759	0822	0845	0912	0942	1012	past	42	12	I		1542	1610	1637	1707	1737	1817	1852	1927	1957	2027
Amersham Station ⊖ ⇌	0803	0826	0849	0917	0947	1017	each	47	17	L		1547	1615	1642	1712	1742	1822	....	1931	....	2031
Chesham Broadway ⊖	0812	0835	0858	0928	0958	1028	hour	58	28			1558	1623	1650	1720	1750	1830	....	1940	....	2040
Chesham Pond Park Estate	0820	0843	0908	0938	1008	1038		08	38			1608	1631	1658	1728	1758	1838	....	1948	....	2048

Saturdays (cont.) / Sundays ★

	Saturdays (cont.)			Sundays ★								
High Wycombe Bus Station	2055	2155	2255	....	1014	1249	1514	1714	....	2114	2314	....
Terriers Cross Roads	2104	2204	2304	....	1023	1258	1523	1723	....	2123	2323	....
Hazlemere Cross Roads	2108	2208	2308	....	1027	1302	1527	1727	....	2127	2327	....
Holmer Green Bat & Ball	2113	2213	2313	....	1032	1307	1532	1732	....	2132	2332	....
Penn Street	2119	2219	2319	....	1038	1313	1538	1738	....	2138	2338	....
Amersham Bus Garage‡	2127	2227	2327	0846	1046	1321	1546	1746	1846	2146	2346	
Amersham Station ⊖ ⇌	2131	2231		0850	1050	1325	1550	1750	1850	2150	2350	
Chesham Broadway ⊖	2140	2240		0859	1059	1334	1559	1759	1859	2159	2359	
Chesham Pond Park Estate	2148	2248		0907	1107	1342	1607	1807	1907	2207	0007	

If you have found this difficult, follow these instructions:

1 Find Amersham Station (look down the list of stops).

2 Run your finger *across* the numbers, until you reach a time *before* 0830. A good time would be 0814.

3 Now look *up* the column from this time, and see what time the bus leaves High Wycombe Bus Station. The time is 0732.

Tasks

1 Look again at the timetable on the opposite page. If you want to catch a train leaving Amersham Station on Saturday at 1030 what time would your bus leave High Wycombe Bus Station? Don't forget to allow plenty of time to buy your ticket!

2 Now, copy this diagram. Use the timetable to fill in the gaps. Find out the time your *whole journey* will take.

A journey from High Wycombe to London

		Time
1	Bus leaves High Wycombe Bus Station	
2	Bus arrives at Amersham Station	
3	Time at station to buy ticket	
4	Train leaves Amersham Station	0935
5	Train arrives at London	1020
	Total journey time	hours minutes

3 Plan a similar journey by bus from your home to your nearest station. You will need a bus *and* train timetable to do this. Work out your total journey time as you have just done above.

4 Look at this *airport arrivals board*, then answer the questions below.

Flight number	Arrival Time	Flight Information	New Arrival Time
BA432	1030	on time	
AI61	0930	delayed by engine trouble	1015
BC612	1035	emergency landing at Manchester Airport	
AF49	0820	delayed due to strong winds	0840

a Which flight is due to arrive at 1015?
b Which flight has the longest delay?
c Which flight had to land unexpectedly?
d You are arriving on flight A161. You hope to catch a coach at 1020. Will you be able to catch it? (Remember to allow at least 20 minutes for landing and baggage collection).

17

Network maps

If you are planning a journey by road, rail or air, there are certain important things you will need to know. These are:

1 What are the *links* between towns and places? For example is there a motorway between Leeds and Bolton? Can you take a train from Edinburgh to Penzance?

2 What are the *stopping places*? These will be stations on a railway line and Service Areas on motorways.

3 Where can you *change routes*? Where are the motorway *intersections* (they always have numbers and are marked on road atlases)? What *junctions* will you need to use to change railway lines? The *key* will tell you which stations are junctions.

Special maps have been designed to help you find out this sort of information. Here is an example of a network map. (Notice that it is very different from the sort of maps you have looked at so far in this book.)

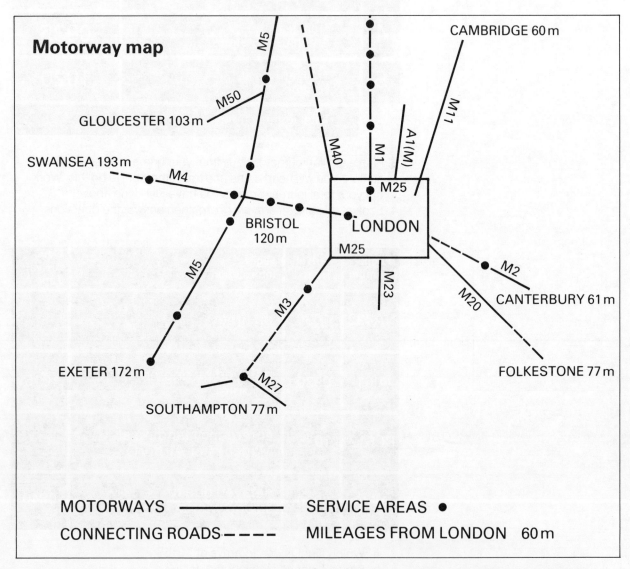

Motorway map

CAMBRIDGE 60 m

M5

M50

GLOUCESTER 103 m

SWANSEA 193 m

M4

M40

M1

A1(M)

M11

M25

BRISTOL
120 m

LONDON

M25

M5

M3

M23

M2

M20

CANTERBURY 61 m

EXETER 172 m

M27

FOLKESTONE 77 m

SOUTHAMPTON 77 m

MOTORWAYS —————— SERVICE AREAS ●

CONNECTING ROADS – – – – MILEAGES FROM LONDON 60 m

Have you spotted that it is not to *scale* (How are distances shown?) Would you describe it as an *accurate* map? Look at the *shape* of the motorways – would they look like this on a real map?

Tasks

1 Look at the motorway map on the opposite page.
 a What is unusual about the *shape* of the M25?
 b Which motorway goes from London to Swansea?
 c How far is it along this motorway from London to Bristol?
 d Which town is 60 miles along the M11 from London?
 e What is unusual about the way the distances from London to Bristol and London to Cambridge are shown on the map?
 f Would you be able to travel on motorway all the way from London to Southampton?
 g Are there any service stations on the M25?
2 Study the map of British Rail InterCity routes in the northern part of the UK. Is there an InterCity link between the following cities?
 a Glasgow and Edinburgh.
 b Newcastle and Carlisle.
 c Harrogate and York.
 d Sheffield and Derby.
3 List the cities you would pass when travelling from Inverness to Preston.
4 Is there a direct link between Blackpool and Carlisle? Newcastle and Edinburgh? Grimsby and Doncaster?

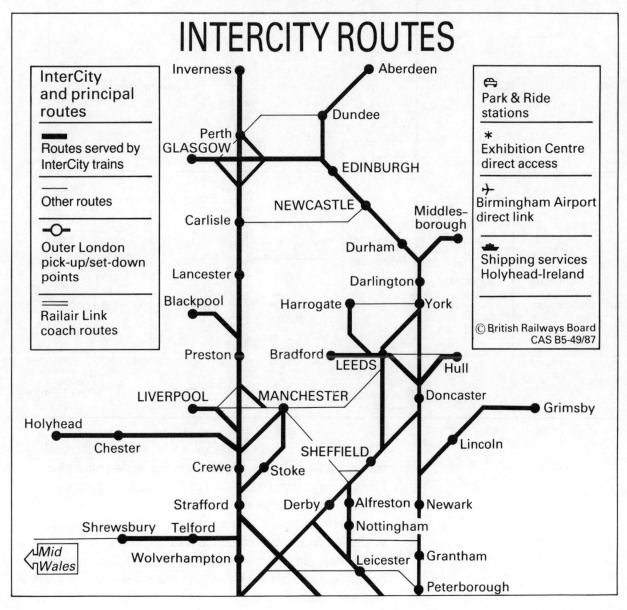

Building and floor plans

When you are visiting a building that you do not know well (perhaps for an interview or to deliver some important mail) it is helpful to use a *building plan*. A plan may be displayed at the entrance to a large complex (such as a shopping or leisure centre). Failing this, the Enquiries Office may be able to give you the information you need.

Plans are very different from the maps you have looked at so far in this book. Look at this plan of a conference centre, which will help you to understand these differences.

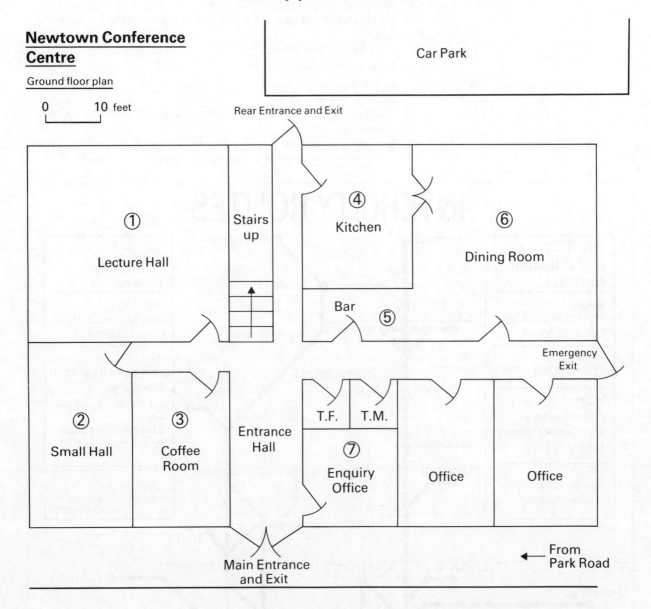

Newtown Conference Centre

Ground floor plan

0   10 feet

1. Plans are drawn to a very large scale, although the scale may not always be shown. What is the scale of this plan?
2. Plans can show you each *floor* of a building or complex. Which floor does this plan show?
3. Plans show details of rooms, offices, hallways and stairs. Which room number is the main dining room?
4. Plans show main entrances and exits as well as access from roads and car parks. Which entrance would you use if you were coming from Park Road to visit the Enquiries Office.

Tasks

Look carefully at this plan of the Open University campus at Milton Keynes. Answer the questions below.

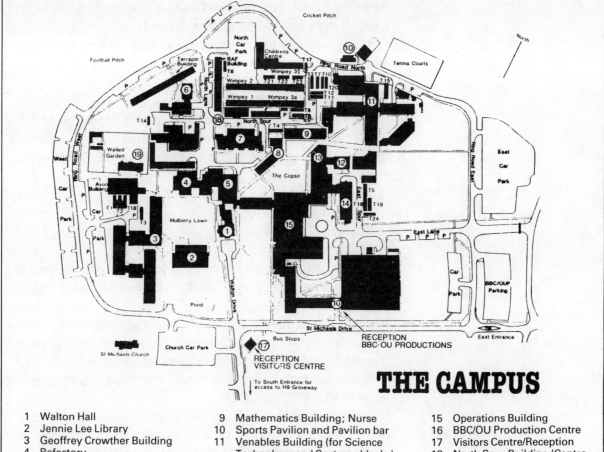

THE CAMPUS

1	Walton Hall	9	Mathematics Building; Nurse	15	Operations Building
2	Jennie Lee Library	10	Sports Pavilion and Pavilion bar	16	BBC/OU Production Centre
3	Geoffrey Crowther Building	11	Venables Building (for Science	17	Visitors Centre/Reception
4	Refectory		Technology and Systems blocks)	18	North Spur Building (Centre
5	Lecture Theatre	12	Earth Sciences		for Continuing Education)
6	Meacham Building (Estates)	13	Computer Centre	19	General Building 1
7	Preparation Laboratories	14	Chambers Building		
8	Academic Computing Service; shop and National Westminster bank				

1 Where would you make an enquiry on arriving at the Campus?
2 What is the name of the nearest car park to the building you have just visited?
3 You wish to visit the Lecture Theatre. What number does it have on the plan?
4 Which road would you walk along to reach the Lecture Theatre from the Visitor's Centre?
5 In which direction would you be walking along this road?
6 You now visit the Refectory. Which area of grass might you be able to see from the windows?
7 Someone you meet in the Refectory asks you if there is parking space outside the North Spur Building. What would be your reply?
8 As you are returning down Walton Drive, a motorist asks you how she can reach the Sports Pavilion. What directions would you give to her? Where would you suggest she parks her car?

Now find a plan of your school or college building.
9 On a copy of the plan, mark where you are now.
10 See if you can mark on the plan your movements around the building today, from the moment you arrived this morning.

UK Locations

Using an atlas

When you want to find out about places or to plan journeys, there are all sorts of *maps* and *atlases* that can help you – you have already used some earlier in this book.

Atlases are especially useful for finding out about *features* and *places* all over the world.

How to use an atlas

The *contents page* at the front of an atlas tells you the *order* in which the maps are arranged. Countries are often listed in *groups*, under headings such as 'Europe' or 'South America'. Here is an example:

    South America
    63 South America relief
    64 South America climate
    65 Brazil

Practise using the contents page of your atlas to find maps of these countries: China (you might need to look under 'Asia'), Scotland and Nigeria.

The *index* at the back of an atlas lists, in alphabetical order, all the places marked on the maps. It will often include rivers, mountains and seas as well as countries and cities.

When you look up a country, you may find an entry like this: Nigeria **92**   8° 3ON   8° OE.

92 is the page number. The other numbers are the *latitude* and *longitude* of Nigeria, which you will need to understand in order to pinpoint *exactly* the place you are looking for.

Latitude and longitude

Lines of latitude (which are always given first) are numbered in degrees and minutes North or South of the Equator. There are 60 minutes in a degree. Place A is 10° North of the Equator. *But* so is place B! So what is the location of place B?

We can pinpoint place B by using lines of longitude, which are numbered West or East of the 'Greenwich Meridian' which, like the Equator is 0°.

You can see that place B is 30° West. So the exact location of place B is 10° ON 30° OW. What is the location of place A?

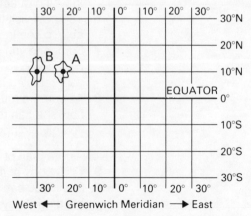

Tasks

Use the contents page of your atlas to find out the following information:

1 a Which page has a world map showing *countries?* (It might be titled 'world; political'.)
   b Look for the Equator on this map and make a list of the countries it passes through.
   c Which country is to the North of the USA on this map?

2 a Which page has a map of Europe showing mountains and rivers? (It might be titled 'Europe; physical'.)
   b List 4 mountain ranges the map shows.
   c List 6 rivers marked on the map.

3 This map of Southern Africa shows 4 important towns, as well as lines of latitude and longitude:

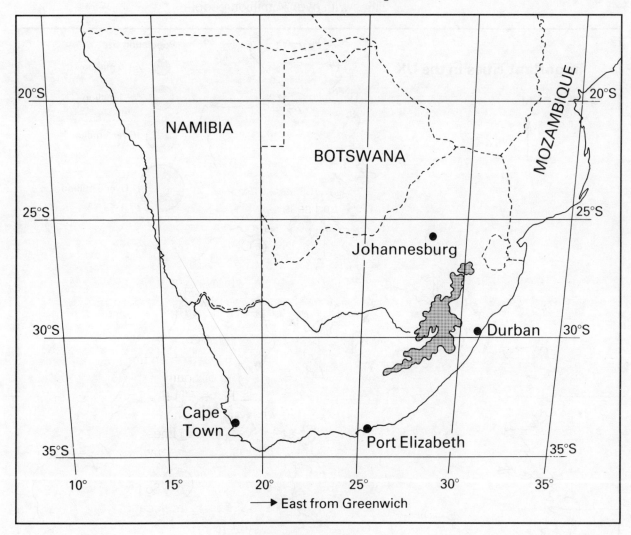

See if you can find each town, using the following information:
a 33° 55S 18° 22E
b 33° 58S 25° 40E
c 29° 49S 31° 1E
d 26° 10S 28° 8E

4 Look up the following cities in the index of your atlas. Find out which *country* each city is in, then, (using the correct page) see if you can locate them on the map: Peking; Montreal; Detroit; Osaka; Leningrad; Delhi.

5 Use the index of your atlas to find out what is unusual about these city names: Perth; Washington; Wakefield; Worcester; Jamestown.

City locations

The map shows the location of 12 of the largest and most important cities in the UK. On your own copy of a map of the UK:

– Mark on each city accurately.
– Use your knowledge, and an atlas if you need to, to name each city (the first letter of each name is given to you).
– Check with the key, and then make lists of the cities, using these headings:
  Cities with more than 5 million people
  Cities with over 1 million people
  Cities with over ½ million people
  Cities with over ¼ million people

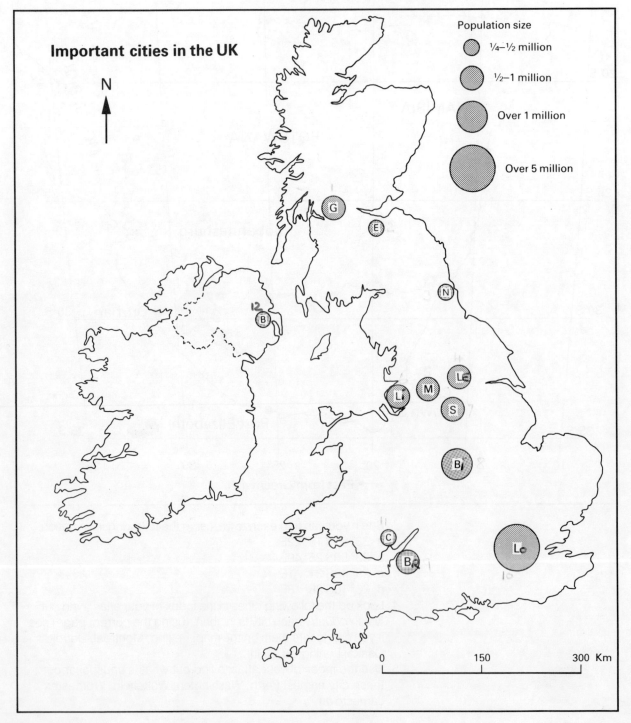

Important cities in the UK

N

Population size

¼–½ million

½–1 million

Over 1 million

Over 5 million

0   150   300 Km

Tasks

1 Each of the sentences below tells you something about one of the cities you have named on your map. From your own knowledge, or by checking with your atlas, match each sentence to the correct city – then write the information on your map next to the correct city.

- The largest city in Northern Ireland, an important port and shipbuilding centre.
- The second largest city in the UK, located in the English Midlands at the centre of the motorway network, and important for motor vehicle and engineering industries.
- The largest city in the South West, an important port and industrial centre.
- The largest city in Wales, an important port and industrial centre.
- The capital city of Scotland, located near the east coast, a major tourist centre, famous for its Festival each year.
- The main industrial city in Scotland, located on the River Clyde.
- Located in Yorkshire, a major industrial city which has important links with the textile industry.
- A major port and industrial city located on the River Mersey in the North West of England.
- The largest city and capital of the UK, a major port and industrial centre, the centre of government, and a very important tourist attraction.
- The largest city in the North West of England, and a major industrial centre.
- The largest city in the North East of England, located on the River Tyne, and an important industrial centre.
- Located near the Yorkshire coalfield, a major industrial city with particular links to the steel and engineering industries.

2 You run a helicopter transport business, based at Newcastle. What is the distance from your business base to the centre of each of the 11 other cities shown on the map?

3 Your firm has taken on the job of making an express delivery to firms in these locations: Belfast; Glasgow; London; Bristol; Liverpool.

   a If all the deliveries are to be made by the same helicopter, make a plan of the shortest route from Newcastle to all those cities, and back to your Newcastle base again.

   b What is the total distance of the route you have planned?

   c Your charges are £1.50 per kilometre. What would the deliveries cost your client?

4 You have decided to expand your helicopter business, and to set up a base at another city further south. Which city would you choose for your new base, and why?

More cities and towns

This map shows Greater London and the Metropolitan Counties of England. Metropolitan Counties are where many people live in towns and cities which have spread across the countryside.

The map also shows some towns and cities in the UK. Remember, there are many more towns that are not shown on the map!

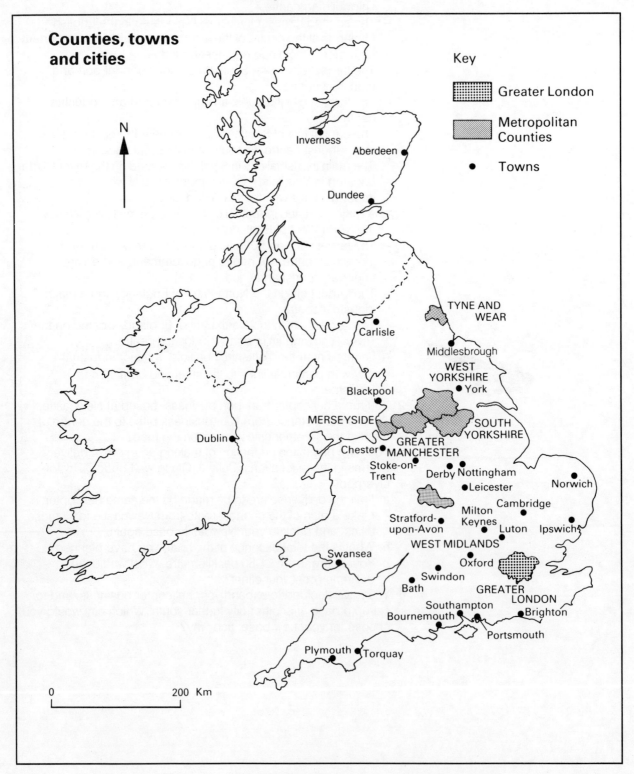

Counties, towns and cities

Key

▓ Greater London

▓ Metropolitan Counties

• Towns

N

Inverness
Aberdeen
Dundee
Carlisle
TYNE AND WEAR
Middlesbrough
WEST YORKSHIRE
Blackpool
York
MERSEYSIDE
SOUTH YORKSHIRE
Dublin
Chester
GREATER MANCHESTER
Stoke-on-Trent
Derby
Nottingham
Leicester
Norwich
Cambridge
Stratford-upon-Avon
Milton Keynes
Luton
Ipswich
WEST MIDLANDS
Swansea
Oxford
GREATER LONDON
Swindon
Bath
Southampton
Brighton
Bournemouth
Portsmouth
Plymouth
Torquay

0   200 Km

On your own copy of a map of the UK, mark in the Counties shown above and, if it is not marked, the county in which you live.

Tasks

1 On page 24 you learned about some of the largest towns and cities in the UK. Use this knowledge to match the following towns with the Metropolitan Counties shown on the opposite page: Newcastle; Birmingham; Liverpool; Leeds; Sheffield.

These questions are all about the towns marked on the map opposite.

2 Five of the towns are located on this map of the Midland Counties. See if you can select the towns and write down their names.

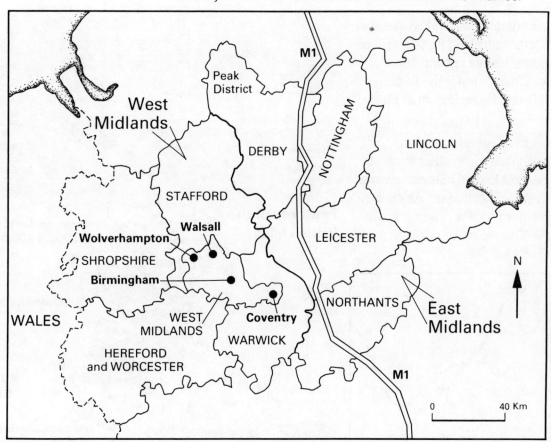

3 Now match each town with one of the following descriptions:
   a This industrial town, a few miles west of the M1, is famous for producing Rolls Royce aero-engines.
   b Many items of hosiery and knitwear are made in this town in the East Midlands.
   c Factories in this town make cigarettes, bicycles and chemical goods, as well as hosiery and knitwear.
   d Much of Britain's pottery (such as dinner and tea sets) is made in this town in the 'Black Country'.
   e Tourists from all over the world visit this town to see where Shakespeare lived.

4 See if you can select the following 'pairs' of towns from the map opposite:
   a Two ancient University towns in England.
   b Two south coast resorts within easy reach of London.
   c Two south coast ports, both near the Isle of Wight.
   d Two popular tourist cities, both famous for ancient and beautiful buildings. One is in the West Country, one is in the North of England.

5 A businessman, whose office is in Derby, asks your travel agency to arrange a trip to Inverness, visiting these towns: Blackpool; Chester; Aberdeen; Carlisle; Edinburgh; Dundee. Put the towns into a suitable order for his journey.

Regions and features

This map shows the land areas of the British Isles. The shaded area is known as the *United Kingdom*.

England, Scotland, Wales and Northern Ireland form the UK as it is generally known. The remaining area is the Republic of Ireland, also known as Eire.

The map below shows the main sea areas around the British Isles as well as some of the best known hills, mountains, rivers and estuaries. An *estuary* is where a larger river (such as the Thames) widens out as it meets the sea.

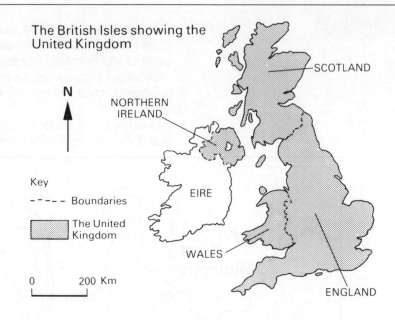

The British Isles showing the United Kingdom

N

SCOTLAND

NORTHERN IRELAND

EIRE

WALES

ENGLAND

Key

– – – – Boundaries

The United Kingdom

0   200 Km

River

Estuary

Sea

Different parts of the UK are often described by using 'compass points'. The boxed area on the map, for example, is often called 'West Country' or the 'South West'. People often talk about the 'Midlands', the 'North East', 'South Wales' or even just the 'North'.

Is the area you live in described in this way?

Copy the features shown on this map on to your own blank map of the British Isles. Colour the high land in brown, and the rivers in blue.

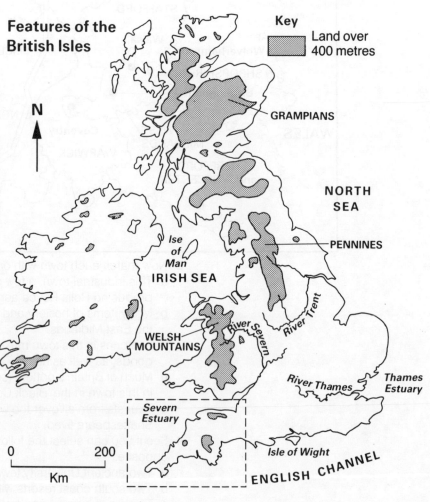

Features of the British Isles

N

Key

Land over 400 metres

GRAMPIANS

NORTH SEA

PENNINES

Ise of Man

IRISH SEA

River Trent

River Severn

WELSH MOUNTAINS

River Thames

Thames Estuary

Severn Estuary

Isle of Wight

ENGLISH CHANNEL

0   200

Km

Tasks

FERRY NEAR MISS
A near disaster happened in the Straits of Dover today as a cross-channel ferry narrowly missed a boat taking part in the Round Britain Boat Race. Dense fog . . .

RACE START DELAYED BY STORM
Appalling weather conditions in the North Sea delayed the start of the Round Britain Boat Race at Lerwick in the Shetland Islands. Twenty foot waves . . .

HUMBER LIFEBOAT RESCUES HOLIDAYMAKERS
The Humber lifeboat was called out today to rescue three men whose boat sank as they watched the Round Britain Race. One of the men . . .

BOY FALLS OFF CLIFF
A family, watching the Round Britain Race from cliffs near Shanklin in the Isle of Wight were horrified to see . . .

HUNDREDS WATCH RACE
Hundreds of people lined the Banks of the Thames Estuary near Southend-on-Sea today to watch the . . .

WINNER LIMPS HOME
The winner of the Round Britain Boat Race limped into Douglas (Isle of Man) harbour today after colliding with another competitor in the Irish Sea, 'Never again', said . . .

FREAK WAVE HITS BOATS – WOMAN MISSING
Rescue boats are still searching for a woman swept overboard by a freak wave in the Bristol Channel today. She was a crew member of . . .

Round Britain Race

Route taken by newspaper race

Event described in cutting

N

0   200
Km

1 This map shows the route taken by a 'Round Britain' power boat race. The newspaper cuttings below describe what happened at various places on the route, which are numbered in flags on the map. See if you can match each headline with the right number on the map. Try to do this without your atlas if you can.

2 A Dutch TV producer, visiting the UK for the first time, will be arriving in Harwich with her car to research suitable locations for a new series. She would like to see the following areas:

a A rugged mountain area where there are ski centres such as Aviemore.

b Another mountain area, named after the highest mountain in Wales.

c A flat part of Eastern England whose large lakes are popular for boating holidays.

d A range of hills and moors, part of which forms the Peak District National Park.

Which part of the UK would you suggest she visits? Which would be the most suitable place for her to begin her tour? Can you suggest a suitable town in each area which she can use as her base for touring?

Britain's motorways

This map shows some of Britain's most important motorways. They were designed to help traffic move easily between cities. There are generally no steep gradients and vehicles can move along without stopping for traffic lights, roundabouts and cross-roads. This is why motorway *intersections* (where motorways meet) and *junctions* (where other roads join motorways) are so complicated.

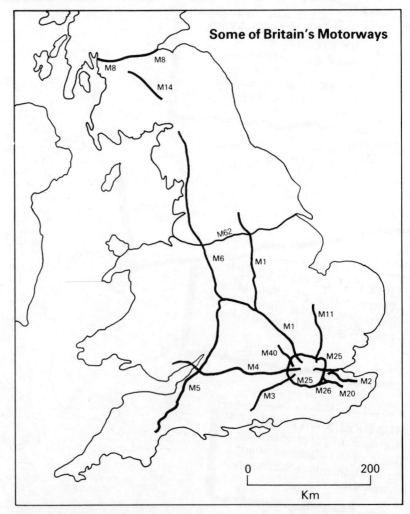

Some of Britain's Motorways

Planning a route

Here is what you need to do when you plan a route using a road atlas. You will also need to remember what you learned on pages 22 and 23.

1 Look at the front of the atlas to find a *Route Planner Map* which will show you the main roads you can use.

2 Find the towns at the beginning and end of your journey. Now you can choose the best route.

3 Find which *pages* of the atlas your route is on, to plan the details of your journey.

4 If you are using a motorway, look for the best junction at which you can start this part of your journey. Also, make a note of the junction number at which you will *leave* the motorway. This is very important, or you may go further than you had intended.

Tasks

1 Local radio stations based in Manchester, Bristol, Leeds, Birmingham, London and Glasgow regularly broadcast traffic news. Referring to the map opposite, suggest which stations would be the first to broadcast these news items?

a High winds caused the Severn Bridge to be closed today to all northbound traffic. There are speed restrictions on large lorries travelling south.

b The busiest holiday weekend of the year has brought traffic to a standstill on the M25 this morning, as thousands of cars try to reach Heathrow Airport.

c A contra-flow system is operating at junction 2 near Coventry on the M1 today. Drivers are advised to allow extra time for their journey.

d A lorry carrying dangerous chemicals overturned near junction 4 on the M8 late last night. The eastbound carriageway is still closed, and drivers are advised to use alternative routes.

e A fresh fall of snow has blocked the M62 east of junction 22. Police say the blockage will take several hours to clear.

2 **a** A sales representative is planning a journey from London to Carlisle, travelling on the M1 and then the M6. Which pages of the road atlas will she use for the last stage of her journey?

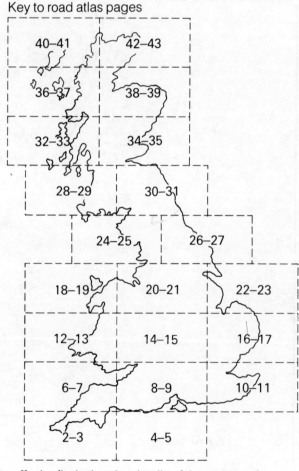

Key to road atlas pages

b As she sets off, she finds that the details of the towns to be visited are in the wrong order. Can you rearrange this list into a suitable order for her to use: Manchester; Stoke on Trent; Coventry; Luton; Birmingham?

3 Can you choose the correct motorway link between these pairs of towns: Birmingham and Bristol; London and Leeds; Hull and Manchester?

More routes and transport centres

This map shows some of Britain's most important ports and airports. Most of the ports are used to shop goods to and from the UK. The freight is either packed into containers (Felixstowe is the largest container port in the UK) or is carried in large lorries. This way of carrying goods is known as Ro-Ro, which is short for 'Roll-on Roll-off'.

Many of the ports also take passengers (and their cars) to European destinations. Some of these destinations are shown on the map on the opposite page.

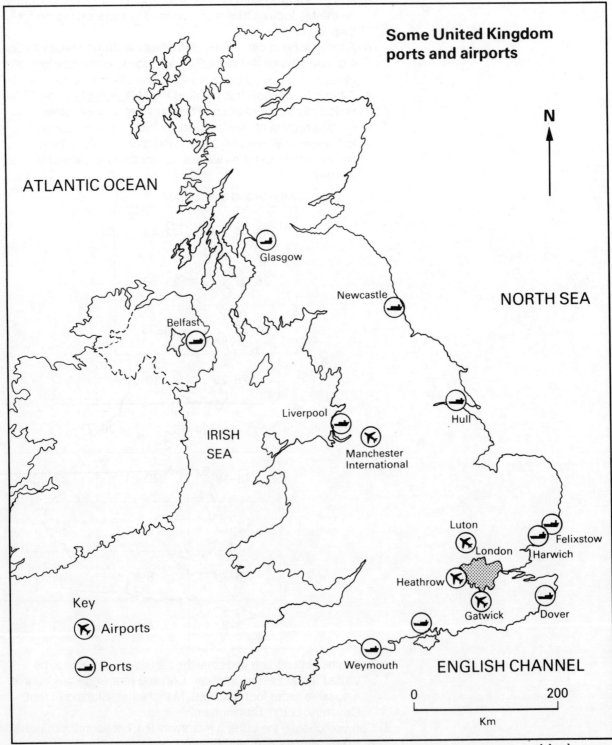

Copy the information shown above on to your own blank map of the United Kingdom.

Tasks

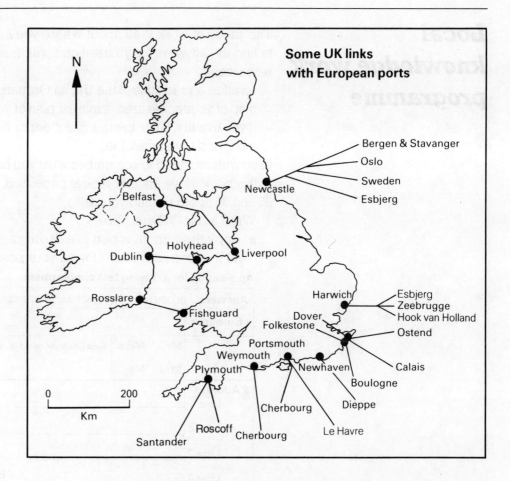

Some UK links with European ports

N

Bergen & Stavanger
Oslo
Sweden
Esbjerg

Newcastle

Belfast

Holyhead
Dublin
Liverpool

Rosslare
Fishguard

Harwich

Esbjerg
Zeebrugge
Hook van Holland

Dover
Folkestone
Portsmouth
Weymouth
Plymouth
Newhaven

Ostend

Calais

Boulogne

Dieppe

Cherbourg

Le Havre

Roscoff
Cherbourg
Santander

0   200
Km

1 The manager of a freight company has to decide which routes to use to export goods to Europe. He then has to fill in a table like the one below. Use the map above, your own knowledge and your atlas to fill in the table for the following goods:

   a Fifty tractors, made in Norwich, to be exported to Denmark.

   b Some engineering equipment, made on Tyneside and destined for Norway.

   c A container load of high fashion winter coats, made in London and to be sold in France.

Goods	City of Origin	British port	Destination (foreign) port
Tractors			
Engineering equipment			
Coats			

2 A fashion company, whose Head Office is in the West End of London, employs your travel agency to arrange journeys for its staff. Where would you advise the following people to begin their journeys? Remember, you will first need to decide whether your clients will need to travel by road, rail, sea or air!

   a The chief designer visiting New York for a fashion show.

   b A sales representative taking a car-load of samples to Paris.

   c The North of England Sales Director, due to attend a midday Board meeting in London.

   d A member of the Advertising Department taking display equipment to a big store in Amsterdam.

Local knowledge work programme

This piece of work is all about *where you live*. You will have to find the answers for yourself and will need to use these resources:

> an atlas; a road atlas of the UK; an Ordnance Survey 1:50000 map of your home area; a map or plan of your town or village: if you live in a large town, a gazetteer or book of street maps: a bus and train timetable.

(You will also need to remember what you have learned earlier in this book. Have another look at pages 6, 8, 10, 12, 16, 22, 24, 26 and 30 just to remind yourself.)

1   Write a title 'Where I live'.

　　a   Copy this form (it is part of a driving test application), and fill it in correctly. Don't forget your *postcode*!

Application for a driving test appointment

Answers	BLOCK LETTERS PLEASE AND WRITE CLEARLY IN INK			
1 Surname				
	Mr　　Miss Mrs　　Ms		First Christian or other name	Other initials
2 Address 　　　Line 1				
Line 2				
Line 3				
Line 4			Postcode	
3 Telephone Number(s)	Home		Work	

　　b   Now use your atlas and a blank map of the British Isles. Find your home town (or nearest large town if you live in a village). Mark it on the map. Add the cities you have already learned about on page 24. Now draw a circle with a radius of 100 miles (put your compass along the scale to do this) around your town. If any of the towns on page 26 are inside the circle, add them to your map. Now list in your book all the large towns that are inside the circle on your map. Add the title 'Large towns within 100 miles of my home'.

　　c   Look in your atlas for a map showing the counties of the UK. Find your county, and make a list of all the counties that are *next* to yours. Give the list a title.

2   Write the title 'Journeys from home'.

　　a   Is there a motorway near your home? If so, write down its number and list some of the towns that can be reached. Find out also which other motorways it joins and list some more important towns that are linked with yours. This list can be given the title 'Towns I can reach by motorway'.

　　b   Is there a station near your home? If so, look at a timetable and find out which towns you can reach by train *without changing*. If you need to change trains to get to other towns and cities, at which station could you do this? Use the title 'Places I can reach by train'.

3 Write a title 'Important places near my home'.

  a Now use your OS 1:50 000 map. Look at this list of buildings, and find one example of each that is near your home. You may need to use the telephone directory to find some, or they may be on your town or village plan. It might be easier to work in a group to do this, as there are a number of places to find.

    1 Hospital which takes emergencies
    2 Police Station
    3 Your Town or County Hall
    4 Your local College
    5 Social Security Office
    6 Public Library
    7 Main Post Office
    8 Leisure Centre
    9 Medical Centre or Doctor's Surgery
    10 Chemist open late on Mondays
    11 Youth Club or Centre
    12 Careers Office

You can make a very useful wall display of these places. First, make 12 different coloured flags. (Use large pins and coloured paper - or you can colour the paper with felt pens.)

  Print a large clear key, which will tell people which building each flag represents.

  Now pin the flags on the map, making sure that they are in *exactly* the right place.

  b Using your local bus timetable and your local map, find out which buses you would use to reach these places and write them down.

  c Make a tape-recording (you may need to write it out first) giving route directions from your home to the nearest Emergency Hospital, firstly by car and secondly by public transport.

  d Talk with your teacher about any other interesting places to find out about in your area (such as tourist sites, big factories or out of town shopping centres). Or think about places that interest you (such as friends' houses, special shops, or local football grounds). Find where *they* are, and make another display map.

4 Now that you have finished this programme, you know much more about your area than some of your fellow students. You could share your knowledge in two ways:

  a Make a big display of all the work you have done.

  b Make a short videotape. Shots of important buildings could be accompanied by short descriptions and verbal directions. You could show the video in the room where your display is on the walls!

World Locations

European countries and cities

The map shows the location and names of 12 countries in Europe. Five great cities are marked and named.

Europe – some countries and cities

Key
- - - - Boundaries of countries
● Cities

ICELAND

NORWAY

NETHERLANDS

DENMARK

UNITED KINGDOM

● Hamburg
● Rotterdam

WEST GERMANY

● Brussels

● Paris

BELGIUM

FRANCE

SWITZERLAND

ITALY

SPAIN

● Rome

GREECE

0   800 Km

On your own copy of this map of Europe:
- Name each country.
- Mark and name each city.
- Choose 3 different colours (or use dark, medium and pale shades of the same colour) to shade in:
  Countries with more than 40 million people (dark colour).
  Countries with between 10 million and 40 million people (medium colour).
  Countries with less than 10 million people (pale colour).

Tasks

1 Look at your map. Which country has the most people? Which has the least? Which do you think is the smallest in size? Which is the largest?

2 Each country you have marked on your map is described in one of the following sentences. Use your own knowledge, and your atlas, to match the description with the correct country:

   a This Scandinavian country is famous for its mountains and fiords: its closest neighbours are Sweden and the USSR.

   b With the Netherlands and Luxemburg, this country forms part of what we call the 'Benelux' countries: many EEC offices are found in its capital city.

   c This country, the largest in Europe, is famous for its food and wine: it is England's nearest continental neighbour.

   d Volcanoes, glaciers and geysers are found in this island in the Atlantic Ocean: many of its people earn their living from fishing.

   e This southern European country was one of the first members of the EEC and many tourists visit its ancient cities and seaside resorts.

   f This small country, almost surrounded by sea, sells much of its bacon, butter and cheese to Britain.

   g This busy industrial country is famous for its iron and steel and chemical industries: the River Rhine passes through it.

   h This unusual, mountainous country has no coastline and is the home of many international organisations such as the Red Cross.

   i With its neighbour, Portugal, this is the most westerly country in Europe: the Costa Brava is one of its popular tourist areas.

   j Many ancient ruins, such as the Parthenon attract visitors from all over the world to this country, which also has many beautiful islands.

   k This group of islands, with many towns and industries, is separated from mainland Europe by a narrow sea.

3 Look at the following list of European air routes and answer the questions below: Hamburg-Athens; Brussels-Rome; London-Rotterdam; Paris-Brussels; Rome-Athens.

   a An American tourist travelling in Europe, wishes to take photographs of the Alps from the air. Which of these routes would you advise her to take?

   b Her friend wishes to visit a cousin who lives in a German port. Which route should she take?

   c Which two cities should they both visit if they are interested in ancient ruins, and do not mind hot weather in the summer?

37

More European countries and cities

This map shows the location and names of 6 more countries in Europe. (The countries you have learned about on pages 36 and 37 have been given numbers).

Five more cities are marked and named. (The cities you have learned about are shown by a dot and the first letter of their name).

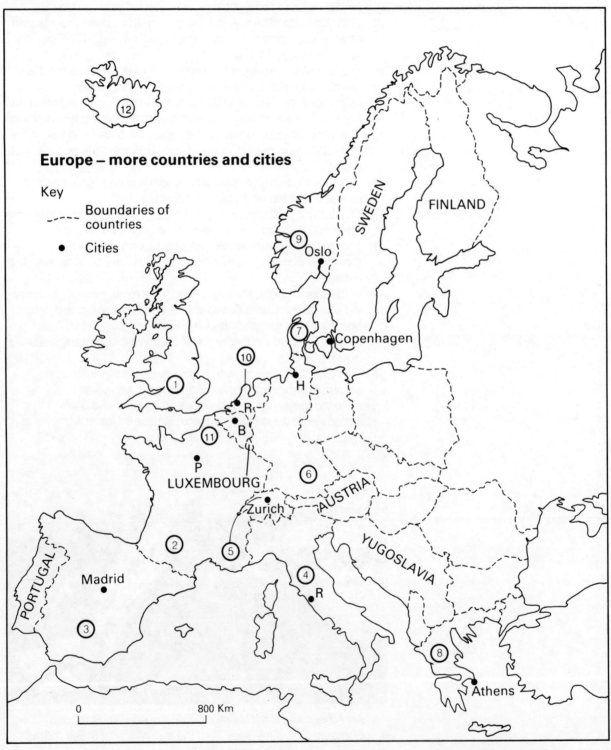

Europe – more countries and cities

Key

- – – – Boundaries of countries
- ● Cities

Make your own copy of a map of Europe marking the countries and cities named in full above. See how many of the countries and cities from pages 36 and 37 you can remember, and mark them on your map as well.

Tasks

1 Each of the sentences below tells you something about one of the countries marked on the map opposite. Use your knowledge and check with your atlas to match each sentence with the correct country.

   a 'Land of a Thousand Lakes'. This country has borders with the USSR, Norway and Sweden.

   b 'A Land of the Midnight Sun'. Stockholm is the capital of this country, which was neutral in World War Two.

   c 'A Princely Principality'. This small, inland country has close links with Belgium and the Netherlands.

   d 'The Land of Port and Sunshine'. The Algarve is a favourite part of this southern European country.

   e 'Home of the Tyrol'. Winter sports are a great attraction in this Alpine country in central Europe.

   f 'A Mediterranean Gem'. Islands, pine trees and mountains are the chief attraction of this country, whose ancient cities (such as Dubrovnik) are popular with tourists worldwide.

2 You are helping to make a series of TV documentaries: which of the cities named on the map opposite would you visit to make each of these programmes?

   a Skiing in Switzerland.

   b The Ancient Ruins of Greece.

   c Bullfighting in Spain today.

   d Land of Mountains and Fiords.

3 Which of the cities you have just named are the capitals of the countries in which they are located?

4 Which of the countries you have named in question 1 are members of the Common Market (EEC)?

Features of Europe

This map shows some of Europe's great rivers and mountain ranges as well as the seas surrounding the coasts. Although the boundaries of the countries are shown, they are not named.

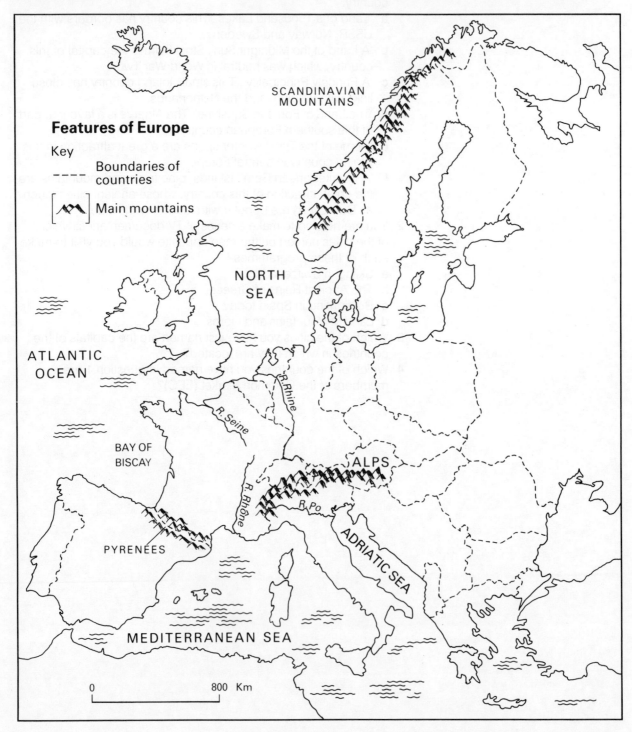

Features of Europe

Key

- - - - Boundaries of countries

Main mountains

SCANDINAVIAN MOUNTAINS

NORTH SEA

ATLANTIC OCEAN

BAY OF BISCAY

R.Seine

R.Rhine

R.Rhône

ALPS

R.Po

PYRENEES

ADRIATIC SEA

MEDITERRANEAN SEA

0   800 Km

The Alps and the Mediterranean coasts are two of Europe's most popular holiday areas. Sixty million people (many of them from northern Europe) are attracted to the hot sunshine and beautiful beaches of the resorts around the Mediterranean Sea. Many also visit the Alps for skiing and other winter sports and to enjoy the beautiful scenery.

On your own copy of the map above see how many of the countries you can name.

Tasks

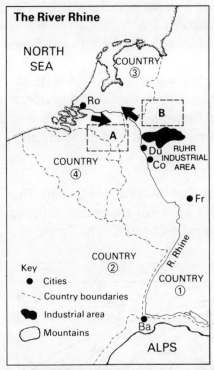

The River Rhine

NORTH SEA

COUNTRY ③

Ro

B

A

COUNTRY ④

Du
RUHR INDUSTRIAL AREA
Co

Fr

R. Rhine

COUNTRY ②

COUNTRY ①

Ba

ALPS

Key
- ● Cities
- --- Country boundaries
- ⬤ Industrial area
- ◯ Mountains

1 This map shows the River Rhine from its source in the Alps to its delta in the North Sea. Use your atlas to name the countries 1, 2, 3 and 4, and the cities Ro, Du, Co, Fr and Ba.

2 How many of these cities are in Germany?

3 Now read this paragraph, which describes the movement of goods on the Rhine, which is the world's busiest waterway:

> 'Coal from the Rhur coalfield is carried downstream to be exported from the great port of Rotterdam. Iron and other raw materials move upstream.

Write down the information you would put in Box A and Box B.

4 Use your knowledge and check with your atlas to answer these questions about holidays in the Alps.
a 'Our company offers Alpine ski holidays in France and Italy as well as Switzerland and southern Germany'. Which important Alpine country has been left out of this list?
b 'Our Head Office in Zurich looks after tourists throughout all our winter resorts'. In which country does this company operate?
c 'Excursions from Chamonix include a spectacular visit to Mont Blanc'. Is this holiday in France or Switzerland?

5 The map below of the western Mediterranean shows some of the airports which are often used for package holidays, as well as some popular resorts.
a What are the mountains marked A on the map?
b Which country has the most airports?
c Which is the most *westerly* country on the map?

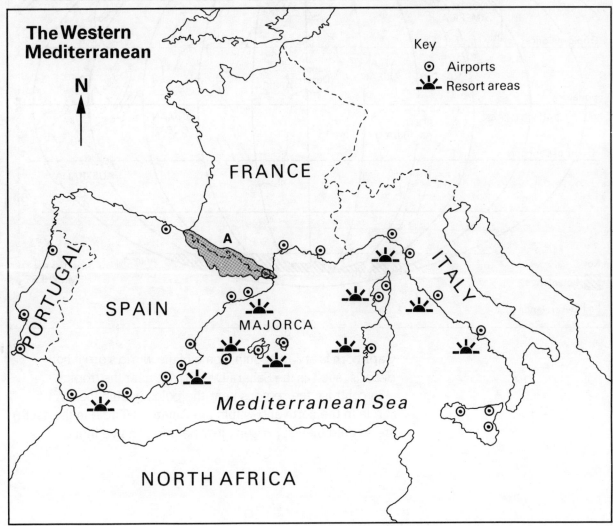

The Western Mediterranean

N

Key
- ⊙ Airports
- ☀ Resort areas

FRANCE

A

PORTUGAL

SPAIN

MAJORCA

ITALY

Mediterranean Sea

NORTH AFRICA

Features of the world

The globe

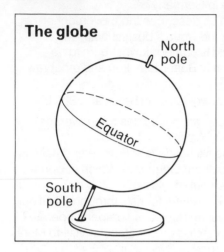

On the opposite page you can see what planet earth looks like from space: much of its surface is hidden by swirling masses of cloud. If we *could* see the world, it would look like a globe – there might be one in your classroom.

The globe shows the *equator* as well as the North and South Poles.

The equator is an imaginary line which circles the earth at its widest point between the poles. It also divides the *Northern hemisphere* from the *Southern hemisphere*.

The map below shows the great oceans, and the main land masses, or continents as they are called.

If you look at the map, you will see the Tropic of Cancer. This is a line of latitude at 23½° North – you may remember learning about latitude and longitude on page 22 and 23. The Tropic of

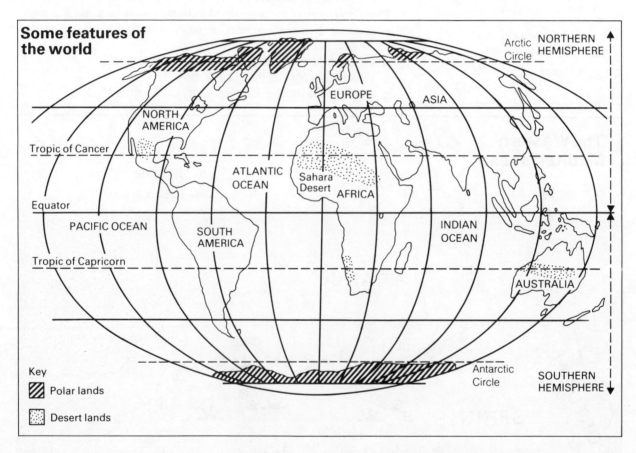

Some features of the world

Key

▨ Polar lands

▦ Desert lands

Capricorn is at 23½° South. Some of the world's great hot deserts (such as the Sahara Desert) are near the tropics.

In contrast, the regions near the poles (within the Arctic Circle in the north and within the Antarctic Circle in the south) are so cold that, at present, few people can live there.

Tasks

Use the information on the opposite page and check with your atlas to answer these questions about the world.

1 Which *ocean* would you cross when travelling between these continents: Africa and South America; Australia and South Africa; North America and Euorope?

2 Which *hemisphere* are these countries in: India; USA; UK; New Zealand?

3 Your firm is thinking of selling its products in Africa and South America: the Marketing Department is collecting information about these two continents. Unfortunately, the files have got mixed up. Which of the files with the titles shown below, would you put in the Africa section of the office?

   Brazilian Forests; Gold from Nigeria; Argentina; Egypt Today; Kenyan Coffee; Chile and the Andes.

4 Sue Watson is Northern Hemisphere Manager of a large group of travel agents, while Joe Martin is Southern Hemisphere Manager. Which of these travellers would Sue's department have to help?
   a A group of tourists held up by a plane strike in Australia.
   b A family who run out of money when visiting Disneyland in the USA.
   c Six students who lose their tickets on a train near Delhi in India.
   d A party of young people whose ski hotel in Switzerland is damaged by an avalanche.

5 Watch tonight's television news, especially the headlines at the beginning. Each time an event takes place *outside* the UK, make a note of it and say *where* the event has taken place. Then, mark each event on a blank map of the world. Do this every night for 4 nights. In which continent have most of the events taken place?

More about the world

This map shows some of the world's great mountains and rivers. Make your own copy of the map.

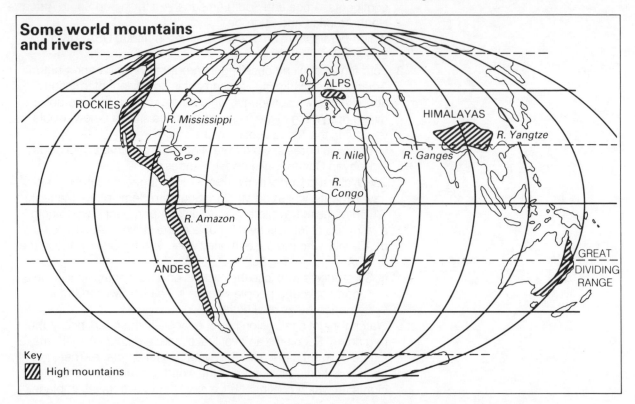

Some world mountains and rivers

ROCKIES

R. Mississippi

ALPS

HIMALAYAS

R. Yangtze

R. Nile   R. Ganges

R. Congo

R. Amazon

ANDES

GREAT DIVIDING RANGE

Key
▨ High mountains

There are many great mountain ranges in the world: one of these is known as the Rockies in North America, and the Andes in South America. Within the range are volcanoes (such as Mount St. Helen's in the USA and Popocatepetl in South America), as well as great basins and deserts (such as the Colorado Desert).

The Himalayas are the highest mountains in the world, with Mount Everest reaching to 8848m.

Some of the world's great rivers are so wide that, at their mouths you cannot see one side from the other. Many are used as a cheap way of transporting goods from one place to another.

Mississippi River

Tasks

1 Each of these sentences describes one of the rivers shown on the map opposite. See if you can name the river in each case.

   a The Pyramids in Egypt are near this great river, whose source is in the mountains many miles to the south.

   b This river in central Africa passes through thousands of square miles of rainforest before reaching the Atlantic Ocean.

   c The rainforest through which this river flows is being cut down to make way for roads and farms. Conservationists are deeply worried about this situation.

   d In the past this great river, much used by junks and other boats, used to flood, drowning many thousands of Chinese people.

   e This river, sacred to many Hindu people, rises in the Himalayas and meets the seas as a large delta, much of which is in Bangladesh.

2 The map below shows the location of National Parks in the Rocky Mountains. Many American teenagers spend their holidays camping in these parks. Which parks do you think match these descriptions in the Camp Company brochure?

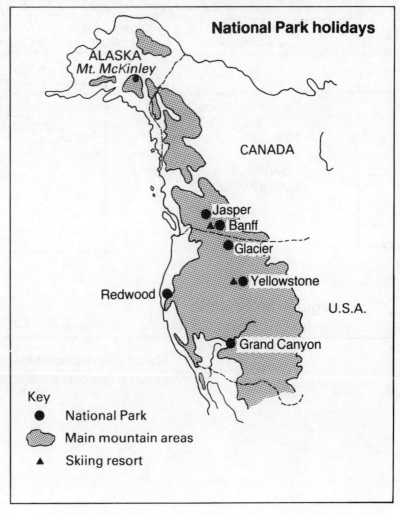

National Park holidays

Key
- ● National Park
- ▨ Main mountain areas
- ▲ Skiing resort

   a 'Come to America's most northerly wilderness: see mountains and glaciers - only for the adventurous (must be over 17 years old).'

   b 'Ski in Canada! Hop over the border and experience hot springs, glaciers and spectacular scenery.'

   c 'Visit the painted desert! Travel by donkey down the world's deepest Canyon! An experience you'll never forget.'

Regions of the world

Sometimes, for a number of reasons, different countries of the world are grouped together and given a 'regional' name. For example, 'Eastern Europe' is used to describe European countries which have communist governments. Television companies will have a 'Far East' correspondent. Sales Organisations may have a 'North American' representative. A travel company will have an 'African' brochure.

This map shows the world's regions, and the names that are sometimes used to describe them.

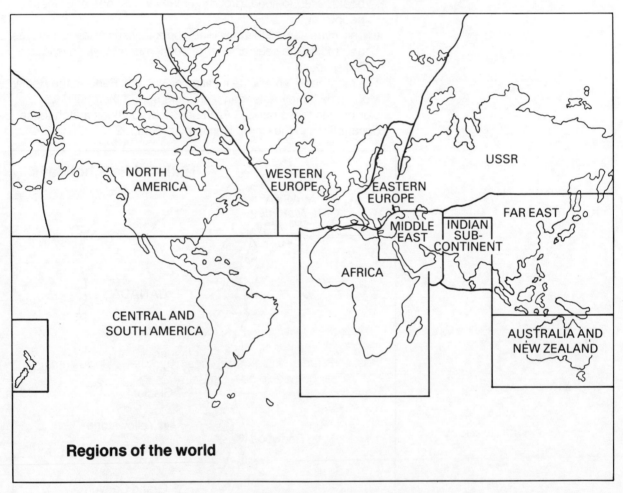

Regions of the world

Mark these regions onto a blank map of the world. Now name the main continents *without* looking at your atlas.

Tasks

1 Use the index of your atlas to locate these countries. Then, using the map opposite say which region each country is in: Japan; Argentina; Nigeria; Pakistan; Egypt; Malaysia.

2 On pages 44 and 45 you learned about some of the world's great rivers. As a reminder, here are their names: Nile; Amazon; Congo; Ganges; Yangtse; Mississippi. Which region is each river in?

3 A soft drinks company has Sales Managers in Australia, the Far East, the Indian sub-continent and Africa. All the managers are competing for a prize, which will be given to the region with the highest sales figures. Here are the sales figures by country (in millions of bottles):

India 6; New Zealand 2; Bangladesh 1; Japan 8; Nigeria 6; Kenya 3; Australia 8; Singapore 2.

Use the table below to fill in the area sales figures. Which manager will win the prize?

Australia		Far East		Indian Sub-Continent		Africa	
Country	Sales	Country	Sales	Country	Sales	Country	Sales
Total		Total		Total		Total	
				The Winner			

4 A TV company has cameramen and reporters throughout Eastern and Western Europe, Russia and the Middle East. The main studios are in Moscow, Paris, Warsaw and Tel Aviv.
   a Can you match each studio with the correct region?
   b Which studio would be responsible for these news items?
      i A car bomb exploding in Beirut.
      ii A plane crash in East Germany.
      iii A World Cup football match in Moscow.
      iv A serious flood in the Dutch coastlands.

World cities

This map shows some of the world's great cities:

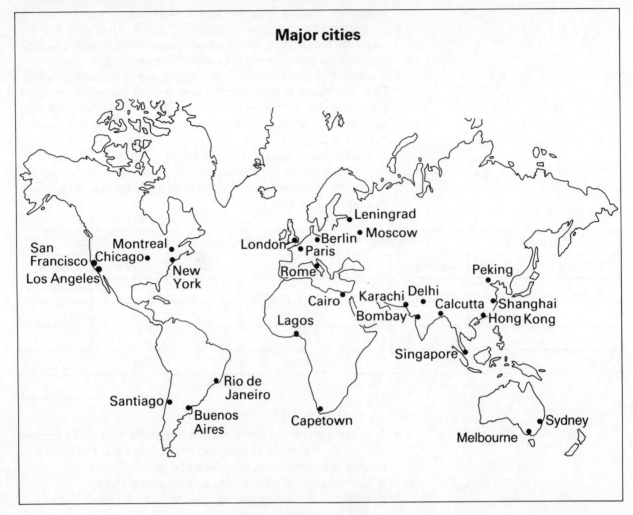

Major cities

Look up the population of these cities in your atlas, to find out which contain *more than* 5 million people.

Now copy the map, showing clearly these largest cities (you will need to use a *key*).

Tasks

1 A wealthy client of your travel agency wishes to travel round the world, visiting the following cities:

New York; Los Angeles; Tokyo; Hong Kong; Singapore; Bombay; Cairo; Moscow; London.

Use the world air routes map below, and your knowledge of world cities to draw his route onto a blank map of the world.

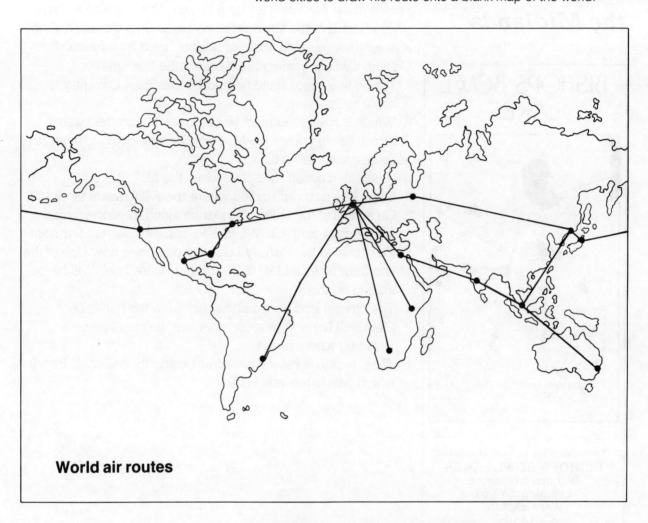

World air routes

Using this map answer these questions:
a Will your client have a chance to fly over the Rockies?
b Will he visit the Southern hemisphere at all?
c At which stage of his journey will he be in the Far East territory of his company?
d At which city will he be nearest to the Himalayas?
e If he decides to change his mind and fly to Peking, at which stage of his journey would it be most convenient to do this?

2 Your client would like some information about the cities he is going to visit. Use your atlas and any other books you may have available to collect this information for him. He would especially like to know:
a The country in which each city is located.
b How many people live in each city.
c Any special tourist features that he could visit.
d What the weather will be like (the journey will take place in January).
e What language most of the people speak.
f In which world region each city is located.

Solving Problems in Geography

Trout fishing in the Midlands

The map below shows the location of Bishops Bowl Lakes, well known in the Midlands for trout fishing. There is also a map giving details of the lakes themselves. Look at the maps, then answer the questions about problems faced by a person from Coventry who is visiting the lakes for the first time.

1. Which two *major* routes would he take from Coventry to drive to Southam?
2. Which B class road must he then take to visit the lakes?

A friend, driving down the M1 from Derby is going to meet him at Southam, via the A426.

3. At which junction should he leave the M1?
4. Should he turn right or left where the A426 meets the A423?
5. On arrival at the Lakes, they drive along the access road towards the car park. Which lake can they see to their right?
6. They stop at the Fishery Lodge to plan their day. One of the two friends would like to boat fish in Mitre Pool. Will he be able to do this?
7. The other friend would like to fish from the banks of Greenhill Lake. Why must he look at the map before deciding where to go?
8. They decide to meet for a picnic beside the waterfall. Beside which lake is the waterfall?

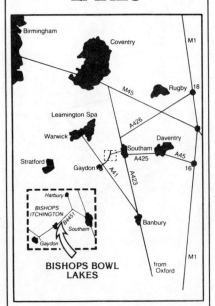

Further information from and enquiries to:
BISHOPS BOWL LAKES
Bishops Itchington
Leamington Spa
Warwickshire
CV33 0SR
Telephone: HARBURY (0926) 613344

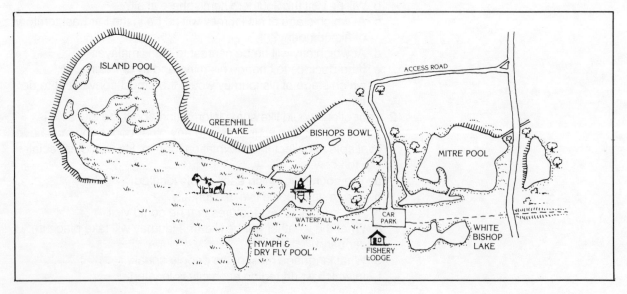

The London Marathon

This map shows the route of the London Marathon. Use your atlas to locate the route on a map of London, and then answer the questions.

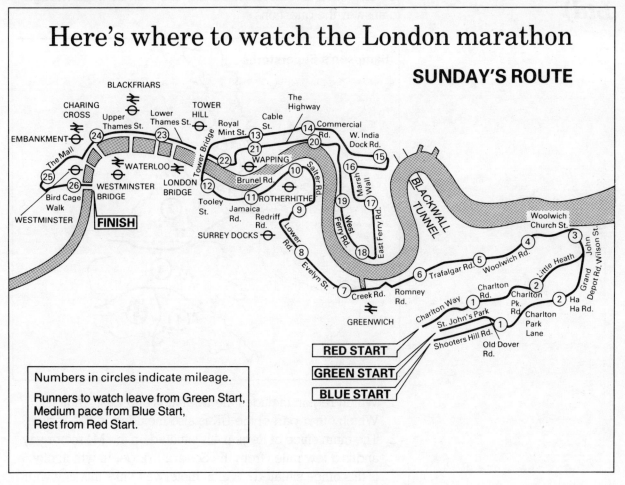

Here's where to watch the London marathon

SUNDAY'S ROUTE

Numbers in circles indicate mileage.

Runners to watch leave from Green Start,
Medium pace from Blue Start,
Rest from Red Start.

RED START

GREEN START

BLUE START

1 Which group of runners would begin their run at Shooters Hill Road?

2 After how many miles do the three starts merge?

3 After how many miles do the runners first cross the Thames?

4 If runners only complete half of the course, would they reach Commercial Road?

5 How many miles of the course are north of the Thames?

6 How many bridges are there between the two crossed by the route?

The police recommend travelling by train:

7 Which station is most likely to be crowded by people arriving to watch the start of the race?

8 Which underground station would be used by someone wishing to cheer a runner after 10 miles of the race?

The route goes near several of London's best known tourist attractions:

9 Which famous buildings connected with the Government of the UK are to be found near Westminster Bridge?

10 One of London's most ancient buildings is about 22 miles along the route. Can you name it?

Sampson's Superstores (UK Ltd)

This map shows the regions of the UK used by a large chain store group, and the location of some of its stores (NB. They are clustered around major cities.) Use your own knowledge to answer the questions.

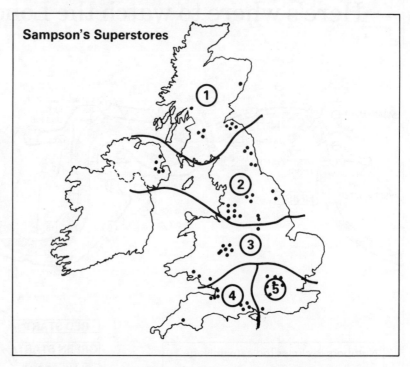

1 Which region includes the Midlands and East Anglia? Which other part of the UK is also included in this region?

2 The main office of Region 4 is situated on the M4 motorway, and is a few miles from the Severn Bridge. In which city is this office situated? Which motorway links this city with Birmingham?

3 Most of the stores in Region 1 are clustered in its two largest cities. Can you name them?

4 Which motorway links the two cities you have named?

5 Which part of the UK forms the most westerly area of Region 2?

6 Which city in this area (well known for its shipyards and textile industries) has the largest cluster of stores?

7 The main warehouse for Region 2 is in Manchester. Which motorway would be used to carry goods eastwards across the Pennines to Hull?

8 Stores in Tyne and Wear are not reaching their sales targets at present: to which region do they belong?

9 Cheese and butter from northern France are shipped to a port in Hampshire. The port is located north west of Portsmouth. Can you name it? Would you estimate that this port is 76 miles, 126 miles or 176 miles from the stores in South London?

10 The company has decided to give each region a name rather than a number. Can you suggest a suitable name for each region?

Cruising in the Indian and Pacific Ocean

Look at this extract from an advertisement for one of Cunard's recent cruises.. Then see if you can answer the questions *without* using your atlas.

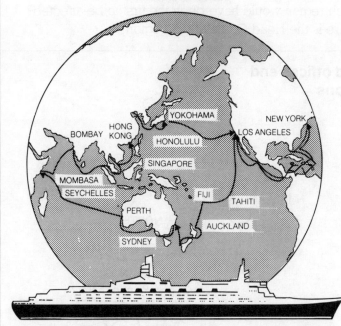

NORMALLY, CUNARD JUST GIVE YOU THE EARTH. THIS WINTER, WE OFFER YOU THE BEST OF BOTH WORLDS.

QE2's PACIFIC AND INDIAN OCEAN CRUISE. SAILS JANUARY 13th

These days, the phrase, 'holiday of a lifetime', is tossed around with rather careless abandon.

But it isn't too difficult to spot the genuine article. And this winter, Cunard don't just offer one genuine example, but two.

QE2 has always been special. Now she's exceptional, and the cruise befits the refitted ship. Instead of spending the winter shivering in Britain, spend it tingling with excitement as one delight follows another.

Her ports of call are a mixture of the exotic, the spectacular, the quaint and the majestic.

There are ports aboard too, and fine wines to match the fine foods. In fact, there's everything it takes to please on board QE2.

Prices vary. For 81 days they start at £9,915. Or, there are many shorter sectors to choose from. 16 days, New York to Los Angeles from £1,920; 36 days, New York to Sydney from £4,390, are just two examples.

If you go on the full cruise:

1. For how long will you be on holiday?
2. What ocean, not mentioned in the advertisement, will you also be crossing?
3. What 3 ports of call will be a real contrast to 'shivering in Britain'?
4. Is the cruise *mainly* in the Northern or the Southern hemisphere?
5. Name one city that you will visit in each of these parts of the world: The Far East; Africa; the Indian Sub-Continent; Australia and New Zealand.
6. Which Japanese city will you visit?
7. Which world famous canal will you sail through in the section between New York and Los Angeles?

If you go on the 36 day sector of the cruise:

8. Which Australian city would you *not* visit?
9. Which New Zealand city *would* you visit?
10. Would you visit Africa on this section of the cruise?

Hurricanes hardly ever happen!

In October 1987 a severe storm, with violent winds passed over Southern England.

This provided a lot of extra work for a Birmingham news agency, which supplies information to newspapers and television networks all over the world. The agency's reporters are based in regions as shown on the map, and each region has a Head Office.

Use what you have learned in this book to answer these questions about the problems the storm caused to the agency and its staff.

First reports suggested that the storm was moving in the direction shown on the map. If these reports were true:

1 Which region would have been the first to be affected?
2 Where is the Head Office of this region?

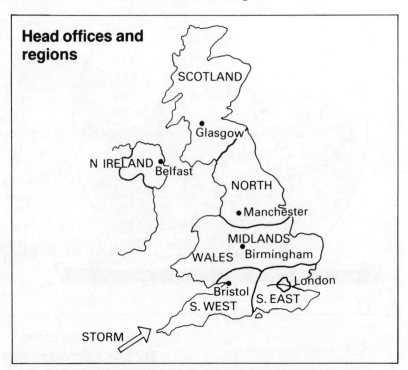

Head offices and regions

Later reports suggested that the South East of England would be severely hit by the storm, so extra reporters would be needed in this area.

3 Use the mileage chart below to suggest which of the regional offices is nearest to London, and should therefore send staff.

B'ham	Bristol	Glasgow	London	Manchester	Penzance
81					
292	373				
105	115	397			
80	161	215	185		
268	185	545	281	342	

As more news of the storm came in, the Graphics Department began to produce a map showing the path of the storm:

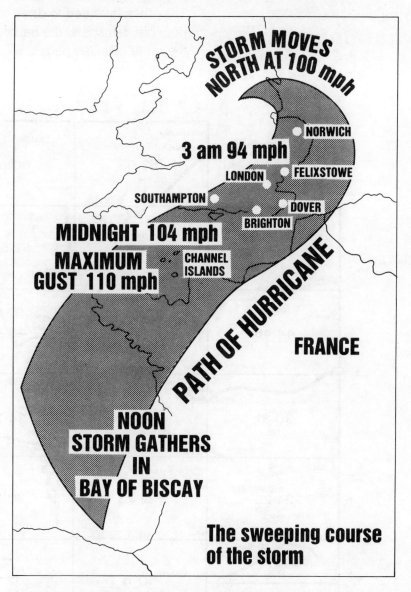

The sweeping course of the storm

An American journalist rang the agency for some information. Can you answer her questions using the map above?

4 Where did the storm begin?
5 Which was the first country the storm passed over?
6 Did the storm pass over the Isle of Wight?
7 Were any cross-channel ports affected?

The morning after the storm the decision was made to send as many reporters as possible to Kent, where the damage was especially bad. The following routes were taken by staff from the Bristol, Birmingham and Manchester offices.

8 Can you suggest which staff used route c?
    a M6, M1.
    b M62, M6, M1.
    c M4.

A group of journalists from Bristol joined the M25 at the M4 intersection, and began travelling in an anticlockwise direction. They then intended to drive to Dover, via the M2, to report on possible damage to the harbour. Their road atlas was open at the route planning page.

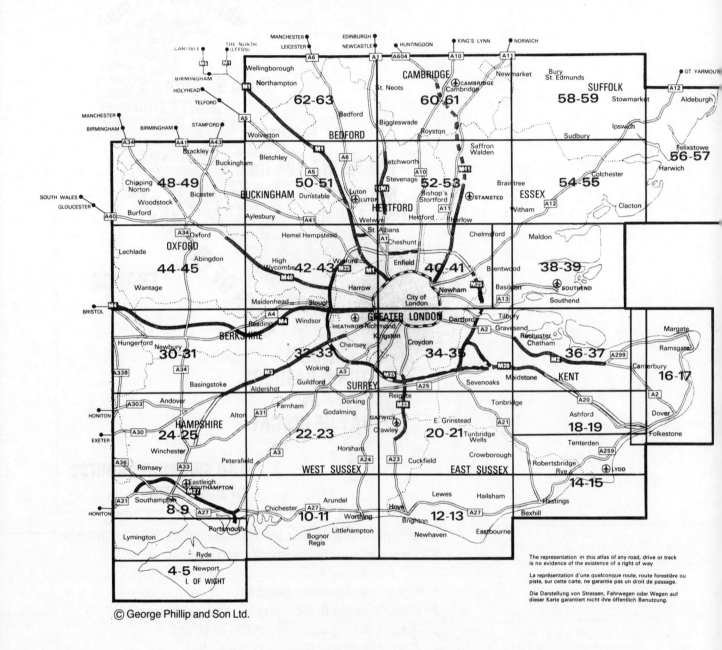

© George Phillip and Son Ltd.

9 List, in order, the pages of the road atlas they would need to use for their journey.

Just as they were passing the M23/M25 intersection, they heard on local radio that a fallen tree had blocked the A2 south of Gravesend.

10 Plan an alternative route to Dover, using major roads. Write a list of the roads, in order. Now list the towns the route passes near.

So many roads were blocked by fallen trees that the journalists had to keep changing their route. Suddenly, they came across a serious accident – a tree had crashed into a house, and firemen were trying to rescue the occupants.

The Senior Fire Officer asked them to fetch a doctor, and gave them a map showing the location of the doctor's house. He told them that the A99 was blocked and that they would be advised to avoid any roads near woodland. This is the map he gave them:

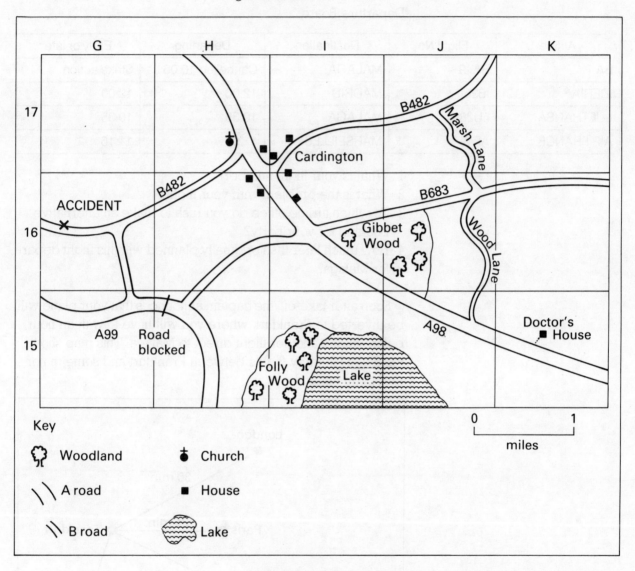

11 In which square of the map did the accident take place?
12 In which square is the doctor's house located?
13 Which two roads meet in the centre of Cardington?
14 Which area of woodland must they avoid when driving to the doctor's house?
15 Use the map, and the information given by the Senior Fire Officer to plan a route between the accident and the doctor's house. Then list the roads in the correct order.
16 Now write down some instructions that could be given to the doctor, telling her how to reach the accident from the centre of Cardington.

Holidaying in Europe

You are to act as courier for a group of young American tourists who will be visiting Spain for the first time. Your job will be to solve any problems connected with their journey and to make sure that they enjoy their holiday as much as possible.

The resort has been chosen, a flight to Malaga has been booked with British Airways (BA) and hotel reservations have been made. Unfortunately, things do not work out as well as you expected!

You arrive at London Airport (Heathrow) and look at the Departure Board:

Airline	Flight No	Destination	Departing	Early or late
BA	BA49	MALAGA	Cancelled 10.06	Strike action
IBERIA	IB264	MADRID	12.06	12.00
LUFTHANSA	LU282	MALAGA	10.35	10.35
AIR FRANCE	AF49	MARSEILLES	11.06	11.16

1 What is your flight number?
2 What is the problem with your flight?
3 To which airline office do you rush to make an alternative booking for your party?
4 How much later than originally planned will this flight depart for Malaga?

Soon after take-off, the captain announces that your plane will be diverted to Frankfurt, where you will have to wait an hour before taking another flight direct to Malaga. This map shows the times taken by flights between Frankfurt and some major European cities:

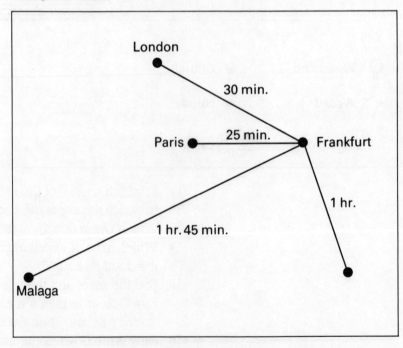

5 Use the information to work out how long your journey between London and Malaga will take.

While you are travelling, you find out more about your party. They are college students and most of them are very keen on swimming and wind surfing. They are all looking forward to visiting the nightclubs in the resort!

On arrival at the airport you contact the tour company office. The representative tells you that, due to overbooking, your hotel is no longer available. He gives you this map of the resort, showing three other hotels which could take your party:

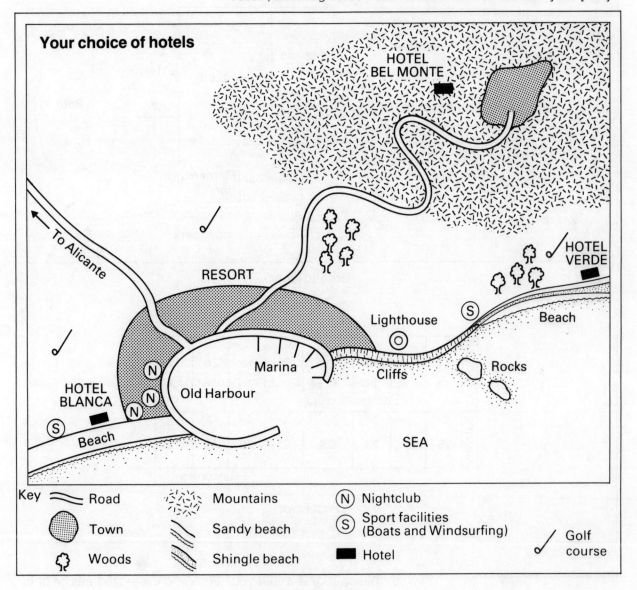

Your choice of hotels

Key:
- Road
- Town
- Woods
- Mountains
- Sandy beach
- Shingle beach
- (N) Nightclub
- (S) Sport facilities (Boats and Windsurfing)
- Hotel
- Golf course

6 Bearing in mind what you have learned about your party, which hotel would you choose for them?

7 Explain why the other two hotels might not be so suitable.

At the hotel, you have to allocate rooms to members of your party. The rooms available are shown on the plan of the first floor shown below:

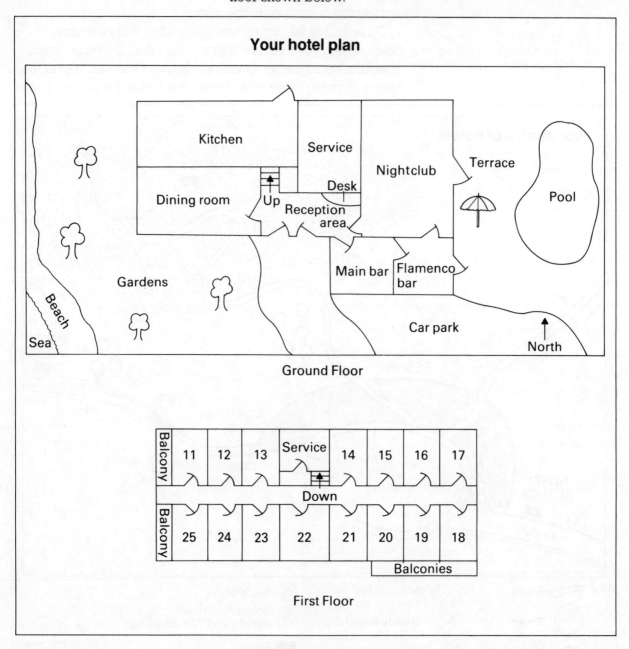

Your hotel plan

Ground Floor

First Floor

8 Which rooms would you allocate to these students, to fit in with requests?

a Karen and Joanne, who want a sea view, and as quiet a room as possible (they do not wish to be near the nightclub!).

b Dave and Mark, who want a south-facing first floor room with a balcony. They don't mind noise!

9 You will need to be as near Reception as possible. Which room would you choose for yourself?

10 A meal has been laid on for your party in the Flamenco Cocktail Bar. How would you tell them where the bar is, in relation to the reception area?

11 The next day, you have to plan 4 outings for your party. They have already given you some ideas as to what they would like to do. Look carefully at the map of the area and suggest a suitable location for each of their suggestions:

a A hike in the mountains, ending in a hill-top village.

b An evening of Spanish dancing, by the floodlit castle walls of an ancient inland town.

c A sand-sailing competition, to be held at the longest beach in this part of Spain.

d Dinner at a harbourside restaurant, near where the main river meets the sea.

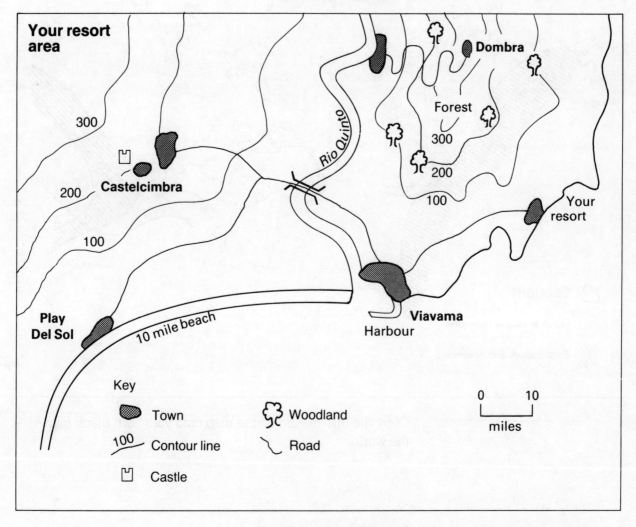

12 Use the scale on the map to work out the distance from your hotel to each of the places you have just named.

13 If the coach company charges a set rate per mile, which outing will cost the most?

World disasters

Much of the work you have done earlier in this book has been about people travelling on holiday, on business or as a part of their daily lives. People who work for international relief organisations have to organise journeys of a different kind. Their job is to move people or supplies to parts of the world where disasters (such as earthquakes) happen, in an attempt to relieve suffering. This map shows parts of the world where such disasters are most likely to happen.

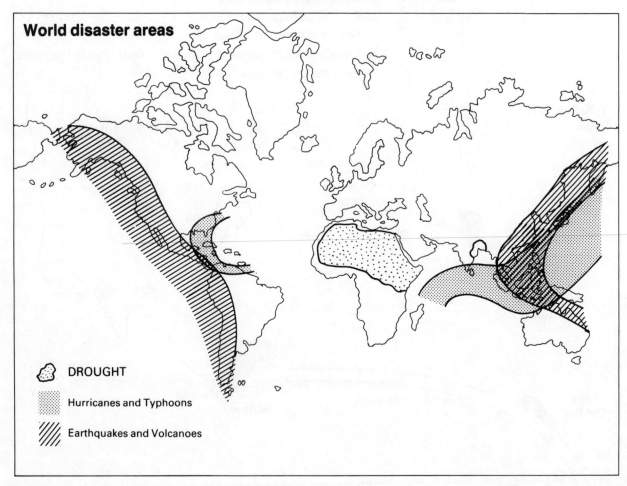

World disaster areas

🌩 DROUGHT

Hurricanes and Typhoons

Earthquakes and Volcanoes

Copy the information on the map onto your own blank map of the world.

Coping with disaster

To do this work, imagine you are working for a relief organisation. It is your job to plan the movement of vital supplies to places where disasters happen. You can work in pairs or small groups, and this is what you will need:

1 A large blank map of the world (A3 would be a suitable size).
2 A way of choosing one number from eight (for example the numbers one to eight could be on small pieces of paper to be drawn from a box).
3 10 red, 10 blue, 10 green, 10 white and 10 black counters (you can make these yourself from coloured card). These represent your supplies as follows:
   – Each red counter = 500 tonnes of food grain.
   – Each blue counter = 200 tents.
   – Each green counter = 50 lorries.
   – Each white counter = 20 packs of medical supplies.
   – Each black counter = specialist staff e.g. doctors or Engineers.
4 An atlas.

Before you start

Name the *world regions* on your map. Now mark and name these *cities*:

London; Calcutta; Santiago; Karachi; Cape Town; Lagos; Mexico City; Hong Kong.

These will be the bases where you store your supplies. Now place your counters on the supply bases. Your specialist staff will be based in London. The map on the previous page will suggest where particular supplies are most likely to be needed.

To begin

1 Draw out a number. This will tell you which of the disasters listed (below) has happened.
2 Mark the position of the disaster as accurately as you can on your map.
3 Decide which is the nearest Supply Base and move the most helpful supplies (counters) to the disaster area.
4 Using your atlas, mark on the map the air, sea or road routes you would use. Remember, your specialist staff will fly from London.
5 Select another disaster and continue until all your supplies have been used.
6 When you have finished, make a list of the supplies that have been sent to each disaster.

The disasters

1 A large area of Bangladesh (50 miles south of Dacca) is flooded as a result of heavy monsoon rains. Thousands are drowned and many acres of land are flooded. Survivors who managed to reach high ground are without food or shelter. Food, medical supplies and tents are needed.

2 An earthquake shatters La Paz, a small town in the Bolivian Andes. Two thousand people are homeless and most buildings in the town are unsafe. An outbreak of typhoid is feared as water supplies are contaminated. Tents, medical supplies and engineering experts are needed.

3 An avalanche engulfs a small village in the Himalayas. Radio reports suggest that this happened north of Delhi and not far from the boundary with Tibet. As yet, nothing else is known.

4 Severe drought, for the third year running, affects people in the Transkei in South Africa. Food is desperately needed as crops have failed.

5 An outbreak of cholera strikes refugee camps in Mali. People living here have fled from the drought-stricken lands on the edge of the Sahara Desert. Medical supplies and doctors are needed.

6 A typhoon devastates the outskirts of Singapore. Many are injured, homes are destroyed and roads are blocked with debris. Tents, engineers, medical supplies and food are needed.

7 An earth tremor is felt in Mexico City, the scene recently of one of the world's most devastating earthquakes. Medical supplies, food, tents and engineers are all needed here.

8 A forest fire is raging near Perth, Australia, after the driest summer in living memory. It is feared that the population may have to be evacuated. Tents are needed.